Current Clinical Neurology offers a wide range of practical resources for clinical neurologists. Providing evidence-based titles covering the full range of neurologic disorders commonly presented in the clinical setting, the Current Clinical Neurology series covers such topics as multiple sclerosis, Parkinson's Disease and nonmotor dysfunction, seizures, Alzheimer's Disease, vascular dementia, sleep disorders, and many others.

Kerry Devlin • Kyurim Kang
Alexander Pantelyat
Editors

Music Therapy and Music-Based Interventions in Neurology

Perspectives on Research and Practice

 Humana Press

Editors
Kerry Devlin
Center for Music and Medicine
Johns Hopkins University School of
Medicine
Baltimore, MD, USA

Kyurim Kang
Center for Music and Medicine
Johns Hopkins University School of
Medicine
Baltimore, MD, USA

Alexander Pantelyat
Center for Music and Medicine
Johns Hopkins University School of
Medicine
Baltimore, MD, USA

ISSN 1559-0585　　　　　　　ISSN 2524-4043　(electronic)
Current Clinical Neurology
ISBN 978-3-031-47091-2　　　ISBN 978-3-031-47092-9　(eBook)
https://doi.org/10.1007/978-3-031-47092-9

© The Editor(s) (if applicable) and The Author(s), under exclusive license to Springer Nature Switzerland AG 2023

This work is subject to copyright. All rights are solely and exclusively licensed by the Publisher, whether the whole or part of the material is concerned, specifically the rights of translation, reprinting, reuse of illustrations, recitation, broadcasting, reproduction on microfilms or in any other physical way, and transmission or information storage and retrieval, electronic adaptation, computer software, or by similar or dissimilar methodology now known or hereafter developed.

The use of general descriptive names, registered names, trademarks, service marks, etc. in this publication does not imply, even in the absence of a specific statement, that such names are exempt from the relevant protective laws and regulations and therefore free for general use.

The publisher, the authors, and the editors are safe to assume that the advice and information in this book are believed to be true and accurate at the date of publication. Neither the publisher nor the authors or the editors give a warranty, expressed or implied, with respect to the material contained herein or for any errors or omissions that may have been made. The publisher remains neutral with regard to jurisdictional claims in published maps and institutional affiliations.

This Humana imprint is published by the registered company Springer Nature Switzerland AG
The registered company address is: Gewerbestrasse 11, 6330 Cham, Switzerland

Paper in this product is recyclable

To all the people who have invited me into their lives during the most intimate moments of theirs. Bearing witness to and being one small part of your stories is the ultimate gift, and I carry them with me always.
—Kerry Devlin

To my dear family members (강기창, 함현숙, 강원석) for their unwavering love and support. Your presence in my life has been a constant source of strength and encouragement, and I am truly blessed to have you by my side. Also, I extend my sincere appreciation to my exceptional work team members, Alex and Kerry, for your outstanding contributions and collaboration. Working alongside you has been a pleasure, and I am grateful for your support, which has enriched our team's dynamics and achievements.
—Kyurim Kang

To my mother Susanna, for inspiring me to become a neurologist and helping me with my first Music Medicine experiment in fourth grade. To my wife Brooke, without whose unwavering and loving support this effort would not have been possible.
—Alexander Pantelyat

Preface

Music Therapy and Music-Based Interventions in Neurology: Perspectives on Research and Practice is intended for music therapy students, practicing music therapists and other clinicians (including, but not limited to neurologists, psychiatrists, psychotherapists, rehabilitation specialists, and nurses) who care for patients with neurological diagnoses. The chapters succinctly cover the key uses of music-based interventions (MBI) for specific neurological diagnoses and associated symptoms. They are written by nearly 40 authors from four continents who have contrasting (yet in our view complementary) expertise. This is intended to encourage the reader to consider multiple approaches in their clinical and research work on MBI. Clinical case vignettes are included throughout to illustrate the various music therapy and music medicine applications in neurology as reviewed in each chapter. Because so much remains to be learned about the mechanisms behind MBI in neurology, future directions for research and clinical work are discussed throughout the book. Whenever possible, key concepts and summary points are combined in tables and figures for the reader's convenience. The editors have endeavored to provide a consistent structural framework for each chapter while maintaining the authors' individual voices. All editors have contributed equally.

Baltimore, MD, USA	Kerry Devlin
Baltimore, MD, USA	Kyurim Kang
Baltimore, MD, USA	Alexander Pantelyat

Acknowledgments

We acknowledge the invaluable contribution of our co-authors in creating this book. It would not have been possible to cover the broad range of Music Therapy and Music Medicine approaches for neurological diagnoses without the complementary expertise of many clinicians, researchers, and educators who contributed to the book's chapters. We are enormously grateful to the patients and families whose collective experience is described in the book's many case vignettes. These vignettes serve as a consistent reminder to the reader of the ultimate purpose of the book: to help improve the quality of care and quality of life for those affected by neurological diagnoses.

Contents

1 **Introduction: Principles and Overview of Music Therapy and Music-Based Interventions** 1
Kerry Devlin, Kyurim Kang, and Alexander Pantelyat

2 **Mechanisms of Music Therapy and Music-Based Interventions** 9
Takako Fujioka and Andrea McGraw Hunt

3 **Music for Stroke Rehabilitation** 23
Anna Palumbo, Soo Ji Kim, and Preeti Raghavan

4 **Music for Traumatic Brain Injury and Impaired Consciousness** 37
Jeanette Tamplin, Janeen Bower, and Sini-Tuuli Siponkoski

5 **Music for Movement Disorders** 49
Yuko Koshimori, Kyurim Kang, Kerry Devlin, and Alexander Pantelyat

6 **Music for Speech Disorders** 71
Yune Sang Lee, Michelle Wilson, and Kathleen M. Howland

7 **Music for Memory Disorders** 85
Hanne Mette Ridder and Concetta Tomaino

8 **Music for Neuro-oncological Disorders** 97
Claudia Vinciguerra, Valerio Nardone, and Matthias Holdhoff

9 **Music Therapy and Music-Based Interventions for Neurologic Palliative Care** 109
Noah Potvin, Maegan Morrow, and Charlotte Pegg

10 **Music for Autoimmune Neurological Disorders** 123
Cindybet Pérez-Martínez, Flor del Cielo Hernández, and Jamie Shegogue

11 **Music for Epilepsy** 137
Robert J. Quon, Ondřej Strýček, Alan B. Ettinger, Michael A. Casey, Ivan Rektor, and Barbara C. Jobst

12	**Music for Surgical/Perioperative Care**........................ 149
	Kelly M. Webber and Myrna Mamaril
13	**Telehealth Music Therapy in Adult Neurological Care**............ 161
	Amy Clements-Cortés and Melissa Mercadal-Brotons
14	**Therapeutic Technology for Music-Based Interventions**........... 173
	Kirsten Smayda and Brian Harris
15	**Music Therapy and Music-Based Approaches with Autistic People: A Neurodiversity Paradigm-Informed Perspective**......... 187
	Hilary Davies and Michael B. Bakan
16	**Psychosocial Aspects of Music Therapy**........................ 199
	Amanda Rosado and Rebecca Vaudreuil
17	**Conclusions and Future Directions**............................ 213
	Kerry Devlin, Kyurim Kang, and Alexander Pantelyat

Appendices.. 217

Index... 229

Contributors

Michael B. Bakan Florida State University, College of Music, Tallahassee, FL, USA

Janeen Bower The Royal Children's Hospital, Melbourne, The University of Melbourne, Parkville, VIC, Australia

Amy Clements-Cortés Music and Health Sciences, University of Toronto, Toronto, ON, Canada

Hilary Davies Guildhall School of Music and Drama, London, UK

Flor del Cielo Hernández Florida State University, Tallahassee, FL, USA

Kerry Devlin Johns Hopkins Center for Music and Medicine, Department of Neurology, Johns Hopkins University School of Medicine, Baltimore, MD, USA

Alan B. Ettinger United Diagnostics and United Medical Monitoring, New Hyde Park, NY, USA

Professional Advisory Board of the EPIC/Epilepsy Foundation of Long Island, East Meadow, NY, USA

Takako Fujioka Center for Computer Research in Music and Acoustics (CCRMA), Department of Music, Stanford University, Stanford, CA, USA

Wu Tsai Neurosciences Institute, Stanford University, Stanford, CA, USA

Brian Harris, MT-BC, NMT/F MedRhythms, Inc., Portland, ME, USA

Matthias Holdhoff Department of Oncology, The Sidney Kimmel Comprehensive Cancer Center at Johns Hopkins, Johns Hopkins University School of Medicine, Baltimore, MD, USA

Kathleen M. Howland Music Therapy and Liberal Arts, Berklee College of Music, Boston, MA, USA

Andrea McGraw Hunt Department of Music, Rowan University, Glassboro, NJ, USA

Barbara C. Jobst Department of Neurology and Neurocritical Care, Dartmouth Health, Lebanon, NH, USA

Geisel School of Medicine at Dartmouth, Hanover, NH, USA

Kyurim Kang Johns Hopkins Center for Music and Medicine, Department of Neurology, Johns Hopkins University School of Medicine, Baltimore, MD, USA

Soo Ji Kim Music Therapy Education Major, Graduate School of Education, Ewha Womans University, Seoul, South Korea

Yuko Koshimori Music and Health Science Research Collaboratory (MaHRC), Faculty of Music, University of Toronto, Toronto, ON, Canada

Yune Sang Lee Department of Speech, Language, and Hearing, School of Behavioral and Brain Sciences, The University of Texas at Dallas, Richardson, TX, USA

Myrna Mamaril Johns Hopkins Hospital, Baltimore, MD, USA

Melissa Mercadal-Brotons Escola Superior de Música de Catalunya, Barcelona, Spain

Maegan Morrow Houston Methodist Hospital, Center for Performing Arts Medicine, Houston, TX, USA

Valerio Nardone Department of Precision Medicine, University of Campania "L. Vanvitelli", Naples, Italy

Anna Palumbo Rehabilitation Sciences Program, Steinhardt School of Education, Culture, and Human Development, New York University, New York, NY, USA

Alexander Pantelyat Johns Hopkins Center for Music and Medicine, Department of Neurology, Johns Hopkins University School of Medicine, Baltimore, MD, USA

Charlotte Pegg Family Pillars Hospice, Bethlehem, PA, USA

Cindybet Pérez-Martínez Washington Adventist University, Takoma Park, MD, USA

Noah Potvin Duquesne University, Pittsburgh, PA, USA

Robert J. Quon Warren Alpert Medical School of Brown University, Providence, RI, USA

Preeti Raghavan Departments of Physical Medicine and Rehabilitation and Neurology, Johns Hopkins University School of Medicine, Baltimore, MD, USA

Ivan Rektor First Department of Neurology, St. Anne's University Hospital and Faculty of Medicine, Masaryk University, Brno, Czech Republic

Multimodal and Functional Neuroimaging Research Group, CEITEC-Central European Institute of Technology, Masaryk University, Brno, Czech Republic

Hanne Mette Ridder Department of Communication and Psychology, Aalborg University, Aalborg, Denmark

Amanda Rosado Independent Author, Germantown, MD, USA

Jamie Shegogue Shenandoah University, Winchester, VA, USA

Sini-Tuuli Siponkoski The University of Helsinki, Helsinki, Finland

Kirsten Smayda, PhD MedRhythms, Inc., Portland, ME, USA

Ondřej Strýček First Department of Neurology, St. Anne's University Hospital and Faculty of Medicine, Masaryk University, Brno, Czech Republic

Multimodal and Functional Neuroimaging Research Group, CEITEC-Central European Institute of Technology, Masaryk University, Brno, Czech Republic

Jeanette Tamplin The University of Melbourne, Austin Health, Melbourne, VIC, Australia

Concetta Tomaino Institute for Music and Neurologic Function, Mount Vernon, NY, USA

Lehman College, City University of New York, Bronx, NY, USA

Rebecca Vaudreuil Henry M. Jackson Foundation for the Advancement of Military Medicine, Inc. in Support of Creative Forces®: NEA Military Healing Arts Network, Bethesda, MD, USA

Claudia Vinciguerra Neurology Unit, University Hospital of Salerno, Salerno, Italy

Kelly M. Webber Medical Cardiac ICU, Mayo Clinic, Rochester, MN, USA

Michelle Wilson Department of Otolaryngology—Head and Neck Surgery, Johns Hopkins Healthcare and Surgery Center, Bethesda, MD, USA

Introduction: Principles and Overview of Music Therapy and Music-Based Interventions

Kerry Devlin, Kyurim Kang, and Alexander Pantelyat

Introduction

While melody and rhythm perception are not unique to humans [1], music is a profoundly human activity. It can uplift us during life's most difficult moments, immerse us in the here and now, inspire us to excel at work or exercise, and drive us to cooperate with others. Traditionally, the study of music has been grouped with the Humanities, but recent decades have seen the flowering of Music Neuroscience. This has been bolstered by the advent of study tools such as functional magnetic resonance imaging and positron emission tomography, which have enabled a better understanding of how our brain perceives music in all its forms [2]. In turn, there has been a burgeoning of research and clinical activity utilizing melody and rhythm to enhance health—in particular, neurological health and wellbeing. In line with increasing public attention to the potential health benefits of music, the National Institutes of Health recently established the Sound Health initiative and its associated funding program, and the music therapy field has been gaining increasing recognition.

This book is intended for those interested in music therapy specifically and the use of music and rhythm generally to improve the lives of those with neurological diagnoses. These are often associated with serious physical and mental health challenges, adversely impact quality of life, and profoundly affect a person's sense of self. Throughout, we aim to embrace a multifaceted approach to this topic and acknowledge the limits (and even potential harms) of reductionism and the traditional medical model when designing music-based interventions (MBI). We also emphasize the need for collaboration between the persons delivering a MBI and

K. Devlin · K. Kang · A. Pantelyat (✉)
Johns Hopkins Center for Music and Medicine, Department of Neurology, Johns Hopkins University School of Medicine, Baltimore, MD, USA
e-mail: kdevlin5@jh.edu; kkang19@jhmi.edu; apantel1@jhmi.edu

© The Author(s), under exclusive license to Springer Nature Switzerland AG 2023
K. Devlin et al. (eds.), *Music Therapy and Music-Based Interventions in Neurology*, Current Clinical Neurology,
https://doi.org/10.1007/978-3-031-47092-9_1

those experiencing it. Notably, the recently published NIH Music-Based Intervention Toolkit [3] acknowledges the fact that *the traditional behavioral/experimental medicine framework is one of several legitimate conceptual frameworks that can be used in MBI research that could merit NIH funding.* Additional conceptual frameworks described in the article include, but are not limited to, a Music Therapy framework (utilizing a combination of psychodynamic, humanistic, behavioral, and music-centered approaches to serve individual needs); a Neuromechanistic framework (conceptual grounding of interventional studies for humans in basic neuroscience research); and a Resilience framework (a context-based support model wherein "therapeutic music environments possess structural elements that support autonomy, encourage the freedom of expression, and promote interaction of patients with their environment") [3]. The authors note that framework choice should be guided by the target patient community, research stage, study design (including specific intervention type, comparison group and outcome measures), and the specific research question.

Who We Are

As an editorial team, we share a collective 30 years of clinical experience working at the intersections of music therapy and music medicine in neurology. We work as a team at the Johns Hopkins Center for Music and Medicine in Baltimore, Maryland, where we are positioned in various roles—senior music therapist and music therapy supervisor (Kerry), postdoctoral research fellow and neurologic music therapist (Kyu), and center director, clinician-scientist and movement disorders neurologist (Alex). We each work with patients with neurological diagnoses and their care partners in outpatient and inpatient contexts and are committed to furthering recognition of the benefits of music therapy and music medicine on both local and global scales.

Book Overview

Our contributing chapter authors are thoughtfully selected experts from different continents (the Americas, Europe, Asia and Australia) and hold different perspectives/worldviews that are sometimes in tension. This was fully intentional: we would like to illustrate the myriad ways MBI can be framed and implemented to benefit those living with neurological diagnoses.

Each chapter includes the self-locations of the authors to help the reader understand the perspective these authors bring to the topic. Relevant terms and concepts are defined and a narrative literature review summarizes the current state of knowledge about research and clinical applications of music therapy and MBI for particular neurological diagnoses. This is supplemented by clinical and community-oriented case vignettes describing relevant applications of the approaches discussed in the chapter. The chapters are rounded out with a conclusions and future considerations section.

Music therapy and music medicine are distinct yet interconnected disciplines that harness the transformative power of music to support physical, emotional, and cognitive well-being. Both music therapy and music medicine make use of MBI. They share a common foundation in recognizing the transformative potential of music (Fig. 1.1), but in our view distinguishing between these enables practitioners, researchers, and patients/clients to make informed decisions, develop targeted interventions, and achieve wide-ranging outcomes in diverse clinical and non-clinical settings.

We use the term "music-based interventions" in this book as a way to refer broadly to the use of music and rhythm to improve human health. Music therapy involves delivery of MBI and/or other therapeutic modalities to individuals or groups by a certified professional with a degree in Music Therapy (minimum Bachelor's degree in the United States, often Master's degree, and at times PhD), with the goal of establishing a therapeutic relationship and supporting well-being in partnership with a patient's larger care team. Music medicine refers to the use of MBI by anyone (including patients themselves) who is not a certified music therapist. Below, we briefly summarize the content of each subsequent book chapter.

We begin by delving into what is known and unknown about the mechanisms of MBI and music therapy as pertains to neurological diagnoses. In Chap. 2, Fujioka and Hunt focus on four areas of inquiry: (1) motor rehabilitation; (2) interpersonal synchrony; (3) pain management; and (4) affect regulation. Activity-dependent neuroplasticity, motor learning, auditory-motor coupling, and basal-ganglia-thalamocortical loop activation are described as key mechanisms relevant to motor neurorehabilitation. The authors also denote four processes underlying interpersonal synchrony: auditory-motor coupling for rhythmic entrainment, empathy processing for understanding others, dopaminergic pathways connecting prediction

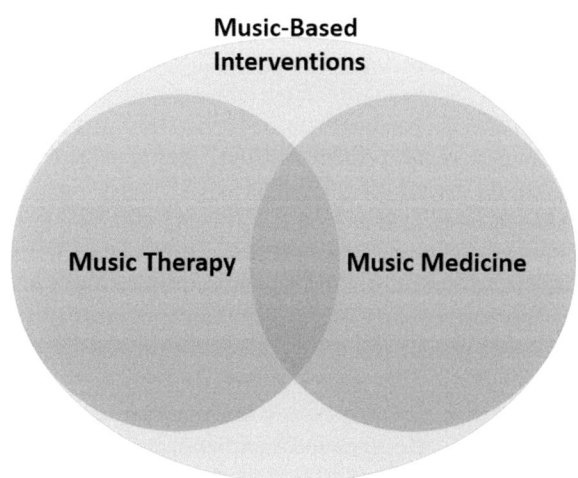

Fig. 1.1 The relationship between music-based interventions, music therapy, and music medicine

and reward and oxytocin's role in social bonding and stress reduction. With respect to acute and chronic pain processing, the authors explain the concept of the neuromatrix and delineate the contributions of the autonomic nervous system (ANS), limbic system, and descending pain modulatory system, Similarly, they describe the roles of the ANS, limbic system and the hypothalamic-pituitary-adrenal axis in affect regulation. In each of these areas, the authors give examples of MBI approaches utilized to date that engage the above mechanisms.

In Chap. 3, Kim, Palumbo and Raghavan summarize the use of music in stroke recovery and rehabilitation. They position the role of MBI, Neurologic Music Therapy (NMT®), and music therapy in the multi-disciplinary context of neurorehabilitation. The impact of music listening on mood, memory and language in those with stroke is linked with structural and functional changes on brain imaging. Rhythmic auditory stimulation and techniques that involve playing musical instruments (including Therapeutic Instrumental Music Performance, Music Supported Therapy and Music Upper Limb Therapy–Integrated) are linked to recovery of motor function. In describing approaches to language recovery following stroke, Melodic Intonation Therapy and group choral singing are highlighted, with the latter approach potentially improving social isolation and enhancing mood. Examples of technology-assisted approaches, such as MusicGlove and musical sonification are provided.

In Chap. 4, Tamplin, Bower and Siponkoski address the use of music for traumatic brain injury (TBI) and disorders of consciousness (DoC). They position MBI as ideal tools for acute TBI recovery because music stimuli can be readily adapted to different levels of arousal and engagement among patients. The authors provide guidelines for delivery of MBI for TBI and DoC, focusing on regulation emotion and arousal, rehabilitation of motor and cognitive function and monitoring and adjusting the MBI over time to support ongoing recovery. The authors conclude by emphasizing that future studies should assess imaging and biofluid markers of TBI and consciousness recovery to delineate the neuroanatomical and neurophysiological correlates of behavioral changes in response to MBI.

The use of MBI for movement disorders is described by Koshimori, Kang, Devlin and Pantelyat in Chap. 5. The authors outline the application of rhythm-based approaches including rhythmic auditory stimulation (RAS®) and other NMT® techniques such as patterned sensory enhancement (PSE®) and therapeutic instrumental music performance (TIMP®) for gait and upper extremity function, with a focus on Parkinson disease. A discussion of integrative music therapy approaches includes experience-oriented and context- or ecologically-oriented music therapy, which are differentiated from, but may be complementary to, functional outcome-oriented paradigms like NMT® [4]. The authors emphasize that a holistic approach to music therapy in movement disorders is more likely to meet the various (and often changing) needs of a patient over time. The authors suggest that future studies should investigate movement disorders such as chorea, dystonia, tics and atypical parkinsonism.

In Chap. 6, Lee, Wilson and Howland provide an overview of MBI for neurological diagnoses affecting speech and language. They describe the connections between language, melody and rhythm processing, discuss the OPERA hypothesis (with an emphasis on brain network overlap for music and language processing) and place this in context of MBI use for developmental language disorders. The authors describe melodic intonation therapy and several NMT® techniques used for speech and language rehabilitation. They conclude with a call for bidirectional collaboration between music therapists, neuroscientists and linguistics experts to improve evidence quality for MBI in the future.

The use of music for memory disorders, with a focus on Alzheimer disease dementia, is described by Ridder and Tomaino in Chap. 7. The authors define different memory types (including musical memory) and describe their neuroanatomical substrates. They explain how music may act as a trigger for memories in the face of neurodegenerative processes. In their literature review, Ridder and Tomaino draw a distinction between the use of music *as* memory within an NMT® framework to music *in* memory (utilizing an "interpersonal, psychotherapeutic approach").

In Chap. 8, Vinciguerra, Nardone and Holdhoff discuss the use of MBI in oncology, focusing on brain cancer, and propose that in the future MBI can be incorporated as part of glioblastoma management at multiple timepoints (Fig. 1.1 in Chap. 8). They describe the adjunctive role of MBI to support patients undergoing surgery (including awake craniotomy), radiotherapy and chemotherapy for cancer and summarize potential benefits of music therapy for reducing anxiety, depression, and pain and improving cognitive function and communication.

Potvin, Morrow and Pegg discuss the use of MBI in neurological palliative care and hospice care in Chap. 9. They describe how MBI, with emphasis on integrative music therapy approaches and NMT® can adapt to shifting quality of life goals for patients and families in this context and evaluate the impact of MBI on overall quality of life, mood, emotional pain and spiritual distress, motor functioning, and enhanced self-efficacy and inclusion. They also stress the importance of considering the role culture plays in the design and implementation of music in health.

In Chap. 10, Pérez-Martínez, del Cielo Hernández, and Shegogue review the use of music for autoimmune disorders affecting the nervous system. They describe frequently occurring neurological symptoms in autoimmune diseases (Fig. 1.1 in Chap. 10) and discuss the importance of evaluating and addressing the patient's intersectionalities (illness, symptoms, social and medical supports, financial aspects, and the overall identity) when designing individualized music-based approaches. A helpful table describes suggested MBI for autoimmune neurological illness grouped by symptom dimension.

In Chap. 11, Quon, Strýček, Ettinger, Rektor, and Jobst discuss the use of music for adults and children with epilepsy. They discuss several hypotheses regarding how music can impact seizures (including activation of the parasympathetic nervous system via the vagus nerve and impact on dopaminergic and GABAergic transmission) and specifically address the use of Mozart's music for seizure reduction. They also summarize the evidence for reduction of interictal epileptiform discharges with MBI in epilepsy. In their view, future research should investigate the effects of MBI and music therapy on common epilepsy comorbidities, including elevated stress, psychobehavioral and cognitive challenges.

In Chap. 12, Webber and Mamaril describe the use of MBI perioperatively. They write about music therapy and music listening approaches (particularly before and after surgical intervention) and emphasize the role of perioperative care nurses in helping patients select personalized music. The role of MBI in reducing self-reported anxiety, pain levels and improving general well-being as well as vital signs is discussed. This is placed in context of attenuation of sympathetic nervous system activity, upregulation of parasympathetic nervous system activity, and reduction of stress levels in both pediatric and adult perioperative care settings.

In Chap. 13, Clements-Cortés and Mercadal-Brotons address the role of Telehealth in delivering Music Therapy to adults with neurological diagnoses. They begin by noting that telehealth music therapy was not widely practiced prior to the COVID-19 pandemic-imposed changes to healthcare delivery. However, because telehealth affords improved access to services for those who may not otherwise be able to receive them (particularly many individuals with neurological diagnoses), the authors point out that telehealth music therapy is here to stay. The authors discuss key challenges for successful telehealth music therapy implementation, including Internet connectivity limitations, delayed sound latency (especially pertinent for music), ensuring digital security and privacy (with ethical considerations), and financial barriers. They cover telehealth NMT®, Music Medicine approaches (vibroacoustic therapy), and virtual community music experiences. The literature review includes the role of telehealth in supporting persons with dementia and their caregivers, as well as those with Parkinson disease, stroke and traumatic brain injury.

In Chap. 14, Smayda and Harris discuss the use of therapeutic technology for MBI. They point out that digital technology enables continuous data collection (including patient-reported outcomes, ecological momentary assessments, and objective physiological measures) that can ultimately lead to individually tailored music interventions (selection of specific musical features) for those with neurological diagnoses such as stroke, Parkinson disease and multiple sclerosis. Table 1.1 in Chap. 14 provides examples of commercial and academic entities developing therapeutic MBI technologies and the literature review highlights several of them. The authors state that for digital MBI technology to be optimally developed and deployed, "it is critical for all parties to work together: music therapists, academic researchers, product developers, and commercial partners."

In Chap. 15, Davies and Bakan provide an overview of the use of music with autistic people using a perspective informed by the neurodiversity paradigm. This approach aims to work *with* (rather than *on*) autistic people to identify goals and build "strengths, confidence, self-acceptance and coping strategies." The authors (a music therapist and an ethnomusicologist) highlight the differences between music therapy- and ethnomusicology-based approaches to the use of music with autistic people and in part ground them in fundamental differences in professional responsibility (wherein the music therapist is often expected to produce "evidence of change through a therapeutic process" and the ethnomusicologist is not). Behavioral music therapy approaches are critiqued as potentially harmful for autistic people's mental health because they often pathologize autistic behavior and aim to mask one's autistic identity. In contrast, community-based music therapy approaches such as sensory-friendly concerts and group singing are described as a positive step in the direction of widespread acceptance of the neurodiversity paradigm [5].

In Chap. 16, Rosado and Vaudreuil discuss the psychosocial aspects of music therapy. They describe the use of composition (rapping, songwriting) as a potential way to improve well-being in underserved communities and music therapy's ability to counteract social isolation (so often experienced in those with neurological diagnoses) and foster a sense of belonging to a community of like-minded people. They review the use of receptive music experiences (music-assisted relaxation, music and imagery, active music listening) to address emotional needs of those with neurological diagnoses in both group and individual contexts. The authors call for future studies that consider psychosocial outcomes as the primary goal, rather than a byproduct, of music interventions.

The concluding Chapter by the book's editors Devlin, Kang and Pantelyat aims to sum up key themes and chart a course for future studies and clinical work. We emphasize the importance of embracing a holistic and collaborative approach to develop and deliver personalized MBI to those with neurological diagnoses. We stress the importance of continued mechanistic investigation to improve understanding of MBI's impact on the nervous system and provide a firm basis for subsequent development of interventions that would promote neuroplasticity and improve functional and quality of life outcomes. We also highlight the importance of keeping a patient's preferences front and center when considering therapeutic approaches that involve music. Finally, we express hope that future work will include patients' voices (describing their lived experience with neurological diagnoses) alongside those of researchers and clinicians.

An Invitation as You Read

Whether sharing music with others or resonating with music shared by someone else, music is, by nature, a shared experience. Given this, we invite anyone reading this book to consider the myriad ways the patient themselves can share in the process of choosing and shaping MBI. As an author team, we believe in the importance of shifting the onus away from a deficit-oriented worldview that locates the therapist/music practitioner as sole expert and patient as "recipient" to one that embraces the patient's own agency and ability to make decisions about how music is most relevant to their own medical care and life context—and why.

In this book, we advocate for a non-binary, multi-faceted approached to clinical music therapy and community music practice that acknowledges the fact that different (and at times, conflicting) approaches *do* exist—and that different patients may require exploration of different approaches to have their needs and desires met in ways most meaningful to them. Where possible, we have intentionally invited contributing authors with differing clinical orientations to co-author chapters in this book as an opportunity to reflect the wide-ranging perspectives about MBI that exist across global clinical and cultural contexts, as well as patient communities. We hope you will find meaning in reflecting on these different ways of conceptualizing the role of MBI in neurology, and that you will consider with us the tension—and possibility—that exist when we acknowledge that there is no such thing as a "right" way to experience music.

References

1. Hoeschele M. Animal pitch perception: melodies and harmonies. Comp Cogn Behav Rev. 2017;12:5–18.
2. Thaut M, Hoemberg V. Handbook of neurologic music therapy. Oxford: Oxford University Press; 2014.
3. Edwards E, St Hillaire-Clarke C, Frankowski DW, Finkelstein R, Cheever T, Chen WG, et al. NIH music-based intervention toolkit: music-based interventions for brain disorders of aging. Neurology. 2023;100:868–78.
4. Bruscia. Defining music therapy. 3rd ed. Barcelona Publishers; 2016.
5. Davies H. 'Autism is a way of being': an 'insider perspective' on neurodiversity, music therapy and social justice. Br J Music Ther. 2022;36(1):16–26.

Mechanisms of Music Therapy and Music-Based Interventions

2

Takako Fujioka and Andrea McGraw Hunt

Introduction

A "mechanism" refers to "a system of causally interacting parts and processes that produce one or more effects" [1]. Specifically, in healthcare, we are concerned with "a biological or behavioral process, the pathophysiology of a disease, or the mechanism of action of an intervention" [2]. Importantly, psychological processes and behaviors can be broken down into simpler steps or actions that are differentiated from strategies, coping skills, and goals [3]. Here, we outline what is known about the major mechanisms of music therapy and music-based interventions (MBI), encompassing motor rehabilitation, social and interpersonal synchrony, pain management, and affect regulation. We review evidence from selected neuroimaging, neurophysiological, neuroendocrine and behavioral studies. We describe the foundations for music-based rehabilitation on the basis of auditory-motor connections, sensorimotor feedback integration, cognitive attention and memory processes. We also consider the mechanisms of music's effect on interpersonal synchrony, including the role of auditory-motor entrainment, empathy processing, the dopaminergic system for prediction and reward, and oxytocin for social bonding and stress reduction (Table 2.1).

T. Fujioka
Center for Computer Research in Music and Acoustics (CCRMA), Department of Music, Stanford University, Stanford, CA, USA

Wu Tsai Neurosciences Institute, Stanford University, Stanford, CA, USA
e-mail: takako@ccrma.stanford.edu

A. M. Hunt (✉)
Department of Music, Rowan University, Glassboro, NJ, USA
e-mail: hunta@rowan.edu

© The Author(s), under exclusive license to Springer Nature Switzerland AG 2023
K. Devlin et al. (eds.), *Music Therapy and Music-Based Interventions in Neurology*, Current Clinical Neurology,
https://doi.org/10.1007/978-3-031-47092-9_2

Table 2.1 MBI goals and key mechanisms

Goal	Key mechanisms	Music-based intervention approach
Motor rehabilitation	Activity-dependent neuroplasticity, motor learning, auditory-motor coupling, basal-ganglia-thalamocortical loop	Music-supported therapy, rhythmic auditory stimulation (RAS®), melodic intonation therapy (MIT)
Social and interpersonal synchrony	Empathy processing, auditory-motor coupling, dopaminergic reward and prediction processing, oxytocin for social bonding and stress reduction,	Rhythmic entrainment, joint music making, singing
Pain management	Neuromatrix, autonomic nervous system (ANS), limbic system, descending pain modulatory system (DPMS)	Active music making, music assisted relaxation, music listening
Affect regulation	ANS, limbic system, hypothalamic-pituitary-adrenal (HPA) axis	Intentional use of familiar/unfamiliar/preferred music and active vs. receptive delivery

Self-Location/Perspectives

We approach this chapter from different, yet complementary, perspectives. Hunt is a music therapy educator and researcher with clinical experience in inpatient psychiatric/residential addictions treatment and private psychotherapy practice. Her research uses behavior and electroencephalography (EEG) to examine neural responses to music therapy interventions. Fujioka, as a cognitive neuroscientist specializing in human non-invasive neurophysiology assessments (magnetoencephalography (MEG)/EEG)) of music cognition, examines issues of neuroplasticity, auditory and motor processing, timing processing, and joint music actions.

Mechanisms of Music in Motor Rehabilitation

Motor rehabilitation relies on the key principle of activity-dependent neuroplasticity, whereby repeated activation of specific neural pathways (involving synchronized firing of groups of neurons) leads to the strengthening of those pathways, based on Hebbian learning principle "neurons wire together if they fire together" described by neuroscientists Löwel and Singer [4]. This fundamental principle implies that lasting brain changes over time develop in response to musical stimulation. It is relevant to diverse brain injury etiologies such as stroke, traumatic brain injury, Parkinson's disease (PD), multiple sclerosis (MS), and other conditions. The use of music for rehabilitation of specific neurological disorders is addressed in other chapters (Chaps. 3–8 and 10). From the motor-learning perspective, the goal of rehabilitation is to obtain generalized motor functionality for various contexts rather than excelling in a fixed, single-motor task [5]. Neurological operations in musical motor skills for auditory-motor connections, sensorimotor feedback integration, and cognitive attention and memory processes provide foundations for rehabilitative benefits.

Hemiparesis in the arm and hand is caused by damage of the corticospinal tracts on the other side (contralateral) of the brain, commonly with stroke (see Chap. 3), spinal cord injury, or brain tumors. One of the most effective physical therapy methods remains constraint-induced motor therapy, which features extensive practice in the paretic hand while restricting the less-affected hand. The former encourages training-induced neuroplastic changes, while the latter suppresses the heavy reliance on the stronger hand and alleviates the inhibitory input from the contralesional hemisphere to the ipsilesional hemisphere [6].

In rehabilitation, incorporating music-based intervention goals of sound making are combined with gross and fine motor skill training in the paretic hand [7–10]. The premotor cortex and supplementary motor area with the prefrontal and parietal association cortices govern the sense of timed actions. Further, the basal-ganglia-thalamo-cortical loop and basal-ganglia-cerebellum loop operate complementarily to plan and fine-tune movement dynamics with cognitive inputs and sensory feedback. Musical rhythm perception and production involve these circuitries (mechanisms of entrainment and synchrony are covered in the next section). The brain also receives the action outcomes through proprioception (sense of muscle states and body position), somatosensation (sense of touch), and vestibular sensation (sense of balance), in addition to visual and auditory sensation, and integrates them in the parietal association areas to support perception-action coupling.

Gait and balance are two major lower extremity concerns, although symptoms and causes vary widely, including dopamine abnormality (PD, Huntington's Disease), focal brain damage (stroke), and nerve fiber damage (multiple sclerosis, Amyotrophic lateral sclerosis). Interventions typically aim to improve gait variability, speed, symmetry, stride, postural stability, and fall resilience. Rhythmic auditory stimulation (RAS®) (see Chap. 5) is one of the most widely used therapeutic approaches for gait training, using an external auditory beat to allow anticipation and timing of each step. RAS® is based on auditory-motor entrainment as described below.

Importantly, one distinct feature of locomotion is the governance provided by the central pattern generator (CPG) in the spinal cord and the brainstem [11], which manages voluntary and automatic controls interactively. Different parts of CPGs organize three key features of walking: rhythm generation, within-leg flexor-extensor muscle coordination, and left-right coordination [12]. The brainstem locomotor regions receive inputs from the premotor cortices, limbic and hypothalamic areas, as well as the parietal lobe, basal ganglia, and the cerebellum to communicate with the spinal cord circuitries for movement planning, initiation and stopping, emotion and sensory information integration, and anticipatory postural controls. Somatosensory signals from skin receptors also contribute to these circuits for detecting and avoiding obstacles.

The motor system communicates through neural oscillatory activities in beta-band (13–30 Hz) [13] observed at the level of local neurons to the macroscopic level such as with human EEG/MEG. They are modulated by rhythm perception [14, 15], the gait cycle [16], and its adaptation [17]. Freezing of gait in Parkinson's disease is linked to exaggerated subthalamic nuclei beta oscillations considered to interfere with the cortico-basal ganglia networks [18, 19]. One of the brainstem locomotor areas, the pedunculopontine nucleus, also shows alpha and beta band synchronization for imagined gait [20], indicating their extended role beyond simple pattern generation. As such, RAS® likely encourages compensatory communications to the brainstem and spinal locomotor areas [21], although details should unfold in future research.

Melodic intonation therapy (MIT) is one of the most successful speech therapies for nonfluent aphasia (see Chaps. 3 and 6). Singing recruits a wider brain area than speaking, including bilateral frontotemporal language and vocal-motor areas with preserved right-hemispheric networks [22, 23]. Tapping with the right hand encourages auditory-motor interaction and articulatory motor functions in the damaged left hemisphere [24]. Despite implications of the name "melodic," rhythmic features of singing rather than pitch may contribute predominantly to MIT's benefits [25, 26].

Mechanisms of Music for Interpersonal Synchrony

For psychosocial benefits, individuals can engage in musical activities together and work towards joint goals by coordinating with each other. Such synchronous actions and associated processes give rise to time-overlapped activities among constituents not only at the behavioural level, but also neural and physiological and affective levels collectively referred as 'interpersonal synchrony' [27], which fosters prosocial behavior, social bonding, and positive affect. Of four major mechanisms underlying interpersonal synchrony, the first two are neuronal processes in (1) auditory-motor coupling for rhythmic entrainment and (2) empathy processing for understanding others in their actions, intentions, and emotions. The remaining two are neuroendocrinological pathways involving (3) dopamine functions to link prediction and reward and (4) oxytocin functions to increase social bonding and reduce stress responses.

Neural Processes: Auditory-Motor Coupling for Rhythmic Entrainment

As described previously, rhythmic entrainment to music and group members involves auditory-motor coupling and timing prediction, engaging the premotor cortex, somatomotor areas, and basal-ganglia-thalamo-cortical circuits working with the cerebellum [28, 29]. These structures generate slow and fast neuronal oscillatory activities synchronized with rhythms and guiding attention [14, 30–32]. Thus, this mechanism is central to supporting a wide range of music-related brain functions.

Neural Processes: Empathy for Understanding Others

Empathy refers to both our knowledge of the internal states of others and our response with sensitivity to their distress and suffering [33]. Collectively, empathy processing is crucial in understanding people's actions and emotional states. The brain achieves this via shared representations between self and others, as someone's behavior and one's own would activate the same areas in the sensorimotor and premotor cortices [34]. Emotional contagions such as mimicking facial expressions or vocalizations also offer an unconscious way to align emotionally with others, involving the inferior frontal gyrus and the inferior parietal lobule [35]. In music listening, brain difficulties in the medial frontal cortex and temporal lobes related to attribution of others' mental state ('theory of mind') occurred only in listeners who thought a human composer made the music, compared to those who listened to the same music but believed it as 'computer-made' [36]. This way, the brain manages an empathetic interpretation of the intent of the musical actions and expressions.

Neuroendocrine Processes: Dopamine for Prediction and Reward

Prediction of musical structures is necessary for participatory engagement, even for listening. Prediction occurs for rhythm, melody, texture, or harmony and mediates the whole musical experience. While the dopaminergic system chiefly mediates movement initiation, the mesolimbic pathway, which transports dopamine across the frontal lobe to the nucleus accumbens (NA) and amygdala, is related to prediction. The NA is also associated with anticipating prediction errors and rewards [37, 38]. Indeed, the perceived reward value of music depends on the functional connectivity strength between NA and the auditory cortex [39]. Mismatching between expectation and surprise, mediated by uncertainty, is linked to musical pleasure, involving the auditory cortex, amygdala, and hippocampus, where the NA reflects the uncertainty [40]. Low dopamine levels can disrupt neural messages and cause inattention, loss of focus, or depression. Music engagement might enhance these neural communications and reward processing.

Neuroendocrine Processes: Oxytocin for Social Bonding and Stress Reduction

Oxytocin is a neuropeptide hormone produced within hypothalamus neurons, acting as a neurotransmitter, but it also is released into the bloodstream via the posterior pituitary gland to peripheral targets. Besides its primary reproductive functions like sex, parturition, and lactation, oxytocin contributes to pair bonding and social touch. Importantly, oxytocin modulates serotonin release, another important hormone/neurotransmitter for affect regulation and socialization [41]. Oxytocin improves synchronous movement [42, 43]. Also, administering oxytocin reduces stress response in the hypothalamic-pituitary-adrenal (HPA) axis via delayed transcription of the gene encoding the corticotropin-releasing factor [44]. Singing increases peripheral oxytocin but solo and choir singing may have differential effects [45, 46]. More research linking music engagement and oxytocin dynamics is needed to help advance our understanding of the underlying mechanisms.

Mechanisms of Music for Pain Management

Pain signals involve both the peripheral and central nervous systems, without any single 'pain' center. Pain experience is modulated by sensory, cognitive, and affective processes simultaneously, where a unitary 'body-self' image is maintained through many modules signaling each other—the concept of the 'neuromatrix' [47]. As such, major mechanisms in pain management involve reward, distraction, and affect regulation, engaging overlapping systems such as the limbic and autonomic nervous system (ANS) and the reward processing network. Music has the ability to simultaneously modulate these systems in flexible ways for managing acute and chronic pain [48, 49].

Listening to familiar, pleasing music can promote a sense of safety and comfort, decreasing amygdala activation and indicating the reduction of perceived threat level. This is accompanied by parasympathetic responses of slower breathing and heart rate [50], which is a common approach to controlling pain. However, researchers have found no evidence that specific genres, musical styles or features have a mechanistic effect on pain relief and/or relaxation response [51]. Importantly, music-based pain relief does not rely on a single target mechanism. For example, the brain's reward processing network is strongly affected by music (e.g., [52]), which can affect pain perception. However, increased dopamine production within the reward processing network alone does not lead to music analgesia [53]. Music for self-expression and/or increasing motivation/arousal is also beneficial for pain relief [54], rather than a relaxation response. Furthermore, music listening (ML) decreases the affective dimension of pain (perceived unpleasantness) but not pain intensity (the sensory/discriminative dimension) [55]. Thus, music analgesic effects seem less relevant for sensory and discriminative processes.

Music for Acute Pain

Acute pain could result from tissue/nerve/inflammatory damage related to disease, injury, or medical procedures. Chapter 12 provides more details on preoperative pain management. A trained therapist can support patients during medical procedures with active listening, singing, or playing instruments. These activities facilitate redirection of perception and increase self-efficacy by stimulating the ANS and limbic systems and help alleviate trauma responses and future mental health problems [56]. Outside of medical procedures, clinicians can provide music assisted relaxation using live or recorded music [56]. Here, therapists synchronize music with the participant's body responses (e.g., pulse, respiration rate) to induce relaxation, also modulating the ANS. Therapists may also incorporate sensory imagery, encouraging the use of individualized coping skills, alongside social support and/or medication [48].

Music for Chronic Pain

Chronic pain such as migraine, back pain, fibromyalgia, and diabetic neuropathy also involves neuroplastic changes in the nervous system. These changes include long-term sensitization of pain receptors and somatosensory system impairment caused by nerve damage or acquired brain injury [57]. As such, chronic pain is accompanied by neurological dysfunction of the descending/"top-down" pain modulatory system (DPMS) comprised by cortical and subcortical brain regions that can facilitate or inhibit pain signal input. Importantly, DPMS normally responds to sensory pain signals to activate the endogenous opiate system and thereby suppress pain. However, in chronic conditions, the default mode network (DMN) in the brain responsible for monitoring background information and introspection, shows altered connectivity with the insular cortex (responsible for internal bodily sensation) proportional to subjective pain perception [58].

Psychosocially, chronic pain patients are often aware that their pain will likely never dissipate completely, and experience ongoing impact in their work, social activities, relationships, self-identity, and sleep [49]. Thus, music-related activities can also alleviate co-occurring anxiety and depression by activating the reward system with dopamine release, and the sensorimotor, executive, and limbic structures that process emotions, brain regions that overlap with the DPMS. Interestingly, pain relief is most effective when patients select music themselves. Furthermore, when patients choose music specifically for pain relief rather than based on familiarity, patients experience greater pain relief [59], suggesting the importance of "top-down" control, or sense of agency.

A recently-proposed model of cognitive mechanisms for music-based relief for chronic pain [60] includes musical integration as a component—where, in optimal conditions, ML provides an absorbing experience, leading to an altered sense of time, and where emotional and pleasurable components of the music interact with emotional and physiological components of pain.

Mechanisms of Music for Affect Regulation

Affect refers to evaluative states involving emotion, arousal, mood, stress response, attitude, and more [3]. Affect regulation (AR) is a self-directed process whereby one modulates one or more aspects of their affective state [3] and potentially for the purpose of attaining a comfortable state of arousal [60]. For AR, music interventions can shift one's attention or interpretation of a situation or help one take action to change the immediate situation, causing an affective response. Research into neural mechanisms related to music interventions for AR is limited but growing [50].

Neural Processes of AR: Limbic System, ANS, and HPA Axis

Affective reactivity primarily involves the limbic system, including the amygdala. Affect perception and control then loops in the frontal lobe regions such as the anterior cingulate cortex, oribitofrontal cortex, and lateral prefrontal cortex. These brain areas not only support cognitive control and monitoring [50] but also music and emotion processing [61]. These brain areas further interact with the ANS and HPA axis, the neuroendocrine systems regulating the body's reaction to internal and external events. The relevant biomarkers include heart rate variability (HRV), cortisol, and oxytocin. HRV indicates the body's cardiac adaptability under physiological stress. Cortisol increases in response to stressful or threatening situations, activated by the HPA axis and relevant neural networks. Oxytocin, as explained in the earlier section of this chapter, could also increase in stressful situations, perhaps to promote prosocial behavior [62].

Music can shift these affect-related mechanisms by either upregulating or downregulating arousal response [50]. Thus, it is important that a trained therapist designs interventions according to patients' individualized needs to prevent risk of potential harm. With appropriate music, the aforementioned frontal brain structures would modulate impulses and regulate emotions, along with the amygdala which assesses emotionally salient information. Then, the amygdala may process new, complex information in the musical experience, regardless of arousal or affective content. It is possible to decrease amygdala activation and arousal through non-threatening, familiar music. However, some therapy contexts could indicate sustaining or increasing amygdala activation to maintain or intensify a client's emotional experience, through unfamiliar, novel music, perhaps with dissonance, frequent harmonic changes/shifts, and/or "unpleasant" sounds.

Active and receptive music experiences correlate with decreased cortisol and perceived stress [46, 63, 64], elevated mood [43], modulation of HRV (increased arousal with active music making, decreased with receptive music experience) [64], and increased oxytocin [46] after recovery from surgery [65].

AR for Neurological/Neurodegenerative Conditions

Injury to or neural degeneration of the frontal lobe directly affects brain structures involved in AR. For example, PD is characterized by progressive motor and neuropsychological impairment with abnormal emotion processing, particularly at early stages of the disease, due to the decreased production of dopamine, a neurotransmitter that, in part, promotes reward anticipation and emotion regulation [66]. For these clinical populations, it is important to regulate patients' mood/mitigate depression to increase motivation to engage in rehabilitation while reducing pain and improving quality of life [67]. Pharmacological use of oxytocin in PD patients has led to increased production of dopamine [68]. Brain activity linked to dopamine production has been linked with pleasant emotional responses to music [69]. Music therefore can have a dual effect of promoting oxytocin and dopamine production, making music interventions seem ideal for AR in PD.

In patients with Alzheimer's Disease, positive mood outcomes associated with using familiar music in receptive (rather than active) music therapy [70] and group music experiences [71] align with previously described amygdala activation and threat detection in AR. It is noteworthy that active music therapy may be too physically and psychologically demanding for participants with comorbid health conditions [70]. Future research could elucidate the demands-related efficacy of different music experiences.

Conclusion and Future Directions

The multiplicity of mechanisms underlying the benefits of musical interventions reflects the multidimensional nature of music. Notably, music is a socially constructed endeavor and the therapist-participant relationship is key to its effectiveness [72]. Thus, the shared music experience inherent to music making and listening is embedded within nesting structures of various mechanisms: arousal and stress response, neuroendocrine changes (oxytocin and/or cortisol), and cortical and subcortical networks. The relational nature of music therapy, along with the therapist's expertise, strengthens the potential benefit of interactive interventions. As such, investigating social neuroscience components such as emotional responses to music in interpersonal brain synchrony [73, 74] and rhythmic motor entrainment behaviors with their neural underpinnings may provide insight into ways music therapy promotes therapeutic interpersonal relationships.

The current literature lacks systematic investigation on individuals' creativity or musicality or inclusion of cross-cultural perspectives due to the heavy reliance on Western music and healthcare frameworks. The latter is imperative given that many cultures may not fully differentiate music from expressive activities such as dance [75]. Other areas still in their infancy include mechanistic investigations of music-based interventions in various types of physical and cognitive developmental conditions, dementia, depression, and addiction treatment.

Music therapy research has been criticized for not adequately defining/identifying mechanisms in the conceptual framework of studies, leading to low-quality reports with unclear or conflicting results. Recent guidelines [76, 77] provide ways for transparent and specific reporting of music intervention research, and the National Institutes of Health recently published a Toolkit on Music-Based Interventions [78]. With such guidance, we hope continuing research will support optimization of music therapy interventions, and lead to improved outcomes.

References

1. Craver C, Tabery J. Mechanisms in science. In: Zalta EN, editor. The Stanford encyclopedia of philosophy. Metaphysics Research Lab, Stanford University; 2019. [cited 2022 Oct 20]. Available from: https://plato.stanford.edu/archives/sum2019/entries/science-mechanisms/.
2. "What is a mechanistic study?" NIH: National Institute of Allergy and Infectious Diseases. [cited 2022 Sep 12]. Available from: https://www.niaid.nih.gov/grants-contracts/what-mechanistic-study.

3. Baltazar M, Saarikallio S. Strategies and mechanisms in musical affect self-regulation: a new model. Music Sci. 2019;23(2):177–95.
4. Löwel S, Singer W. Selection of intrinsic horizontal connections in the visual cortex by correlated neuronal activity. Science. 1992;255(5041):209–12.
5. Krakauer JW. Motor learning: its relevance to stroke recovery and neurorehabilitation. Curr Opin Neurol. 2006;19(1):84–90.
6. Taub E, Uswatte G, Elbert T. New treatments in neurorehabiliation founded on basic research. Nat Rev Neurosci. 2002;3(3):228–36.
7. Schneider S, Schönle PW, Altenmüller E, Münte TF. Using musical instruments to improve motor skill recovery following a stroke. J Neurol. 2007;254(10):1339–46.
8. Tong Y, Forreider B, Sun X, Geng X, Zhang W, Du H, et al. Music-supported therapy (MST) in improving post-stroke patients' upper-limb motor function: a randomised controlled pilot study. Neurol Res. 2015;37(5):434–40.
9. Grau-Sánchez J, Duarte E, Ramos-Escobar N, Sierpowska J, Rueda N, Redón S, et al. Music-supported therapy in the rehabilitation of subacute stroke patients: a randomized controlled trial. Ann N Y Acad Sci. 2018;1423(1):318–28.
10. Fujioka T, Dawson DR, Wright R, Honjo K, Chen JL, Chen JJ, et al. The effects of music-supported therapy on motor, cognitive, and psychosocial functions in chronic stroke. Ann N Y Acad Sci. 2018;1423(1):264–74.
11. Kiehn O. Locomotor circuits in the mammalian spinal cord. Annu Rev Neurosci. 2006;29:279–306.
12. Takakusaki K. Neurophysiology of gait: from the spinal cord to the frontal lobe. Mov Disord. 2013;28(11):1483–91.
13. Engel AK, Fries P. Beta-band oscillations-signalling the status quo? Curr Opin Neurobiol. 2010;20(2):156–65.
14. Fujioka T, Trainor LJ, Large EW, Ross B. Beta and gamma rhythms in human auditory cortex during musical beat processing. Ann N Y Acad Sci. 2009;1169:89–92.
15. Fujioka T, Trainor LJ, Large EW, Ross B. Internalized timing of isochronous sounds is represented in neuromagnetic beta oscillations. J Neurosci. 2012;32:1791–802.
16. Gwin JT, Gramann K, Makeig S, Ferris DP. Electrocortical activity is coupled to gait cycle phase during treadmill walking. NeuroImage. 2011;54(2):1289–96.
17. Wagner J, Makeig S, Gola M, Neuper C, Müller-Putz G. Distinct β band oscillatory networks subserving motor and cognitive control during gait adaptation. J Neurosci. 2016;36(7):2212–26.
18. Singh A, Plate A, Kammermeier S, Mehrkens JH, Ilmberger J, Bötzel K. Freezing of gait-related oscillatory activity in the human subthalamic nucleus. Basal Ganglia. 2013;3(1):25–32.
19. Toledo JB, López-Azcárate J, Garcia-Garcia D, Guridi J, Valencia M, Artieda J, et al. High beta activity in the subthalamic nucleus and freezing of gait in Parkinson's disease. Neurobiol Dis. 2014;64:60–5.
20. Lau B, Welter M, Belaid H, Fernandez Vidal S, Bardinet E, Grabli D, et al. The integrative role of the pedunculopontine nucleus in human gait. Brain. 2015;138(5):1284–96.
21. Damm L, Varoqui D, De Cock VC, Dalla Bella S, Bardy B. Why do we move to the beat? A multi-scale approach, from physical principles to brain dynamics. Neurosci Biobehav Rev. 2020;112:553–84.
22. Zarate JM. The neural control of singing. Front Hum Neurosci. 2013;7:237.
23. Wan CY, Zheng X, Marchina S, Norton A, Schlaug G. Intensive therapy induces contralateral white matter changes in chronic stroke patients with Broca's aphasia. Brain Lang. 2014;136:1–7.
24. Schlaug G, Norton A, Marchina S, Zipse L, Wan CY. From singing to speaking: facilitating recovery from nonfluent aphasia. Future Neurol. 2010;5(5):657–65.
25. Stahl B, Kotz SA, Henseler I, Turner R, Geyer S. Rhythm in disguise: why singing may not hold the key to recovery from aphasia. Brain. 2011;134(10):3083–93.
26. Zumbansen A, Peretz I, Hébert S. The combination of rhythm and pitch can account for the beneficial effect of melodic intonation therapy on connected speech improvements in Broca's aphasia. Front Hum Neurosci. 2014;8:592.

27. Rennung M, Göritz AS. Prosocial consequences of interpersonal synchrony: a meta-analysis. Z Psychol. 2016;224(3):168–89.
28. Grahn JA, Brett M. Rhythm and beat perception in motor areas of the brain. J Cogn Neurosci. 2007;19(5):893–906.
29. Chen JL, Penhune VB, Zatorre RJ. Listening to musical rhythms recruits motor regions of the brain. Cereb Cortex. 2008;18(12):2844–54.
30. Lakatos P, Gross J, Thut G. A new unifying account of the roles of neuronal entrainment. Curr Biol. 2019;29(18):R890–905.
31. Henry MJ, Obleser J. Frequency modulation entrains slow neural oscillations and optimizes human listening behavior. Proc Natl Acad Sci U S A. 2012;109(49):20095–100.
32. Fujioka T, Ross B, Trainor LJ. Beta-band oscillations represent auditory beat and its metrical hierarchy in perception and imagery. J Neurosci. 2015;35(45):15187–98.
33. Batson CD. These things called empathy: eight related but distinct phenomena. In: Decety J, Ickes W, editors. The social neuroscience of empathy. online edn. Cambridge, MA: MIT Press Scholarship Online; 2009, 22 Aug 2013.
34. Decety J, Jackson PL. The functional architecture of human empathy. Behav Cogn Neurosci Rev. 2004;3(2):71–100.
35. Shamay-Tsoory SG. The neural bases for empathy. Neuroscientist. 2011;17(1):18–24.
36. Steinbeis N, Koelsch S. Understanding the intentions behind man-made products elicits neural activity in areas dedicated to mental state attribution. Cereb Cortex. 2009;19(3):619–23.
37. McClure SM, Berns GS, Montague PR. Temporal prediction errors in a passive learning task activate human striatum. Neuron. 2003;38(2):339–46.
38. Pessiglione M, Seymour B, Flandin G, Dolan RJ, Frith CD. Dopamine-dependent prediction errors underpin reward-seeking behaviour in humans. Nature. 2006;442(7106):1042–5.
39. Salimpoor VN, Van Den Bosch I, Kovacevic N, McIntosh AR, Dagher A, Zatorre RJ. Interactions between the nucleus accumbens and auditory cortices predict music reward value. Science. 2013;340(6129):216–9.
40. Cheung VKM, Harrison PMC, Meyer L, Pearce MT, Haynes J, Koelsch S. Uncertainty and surprise jointly predict musical pleasure and amygdala, hippocampus, and auditory cortex activity. Curr Biol. 2019;29(23):4084–4092.e4.
41. Mottolese R, Redouté J, Costes N, Le Bars D, Sirigu A. Switching brain serotonin with oxytocin. Proc Natl Acad Sci U S A. 2014;111(23):8637–42.
42. Gebauer L, Witek MAG, Hansen NC, Thomas J, Konvalinka I, Vuust P. Oxytocin improves synchronisation in leader-follower interaction. Sci Rep. 2016;6:38416.
43. Josef L, Goldstein P, Mayseless N, Ayalon L, Shamay-Tsoory SG. The oxytocinergic system mediates synchronized interpersonal movement during dance. Sci Rep. 2019;9(1):1894.
44. Jurek B, Slattery DA, Hiraoka Y, Liu Y, Nishimori K, Aguilera G, et al. Oxytocin regulates stress-induced Crf gene transcription through creb-regulated transcription coactivator 3. J Neurosci. 2015;35(35):12248–60.
45. Schladt TM, Nordmann GC, Emilius R, Kudielka BM, de Jong TR, Neumann ID. Choir versus solo singing: effects on mood, and salivary oxytocin and cortisol concentrations. Front Hum Neurosci. 2017;11:430.
46. Good A, Russo FA. Changes in mood, oxytocin, and cortisol following group and individual singing: a pilot study. Psychol Music. 2021;50(4):1340–7.
47. Melzack R, Katz J. Pain. WIREs Cogn Sci. 2013;4(1):1–15.
48. Loewy J. Underlying music mechanisms influencing the neurology of pain: an integrative model. Brain Sci. 2022;12(10):1317.
49. Gold A, Clare A. An exploration of music listening in chronic pain. Psychol Music. 2013;41(5):545–64.
50. Moore KS. A systematic review on the neural effects of music on emotion regulation: implications for music therapy practice. J Music Ther. 2013;50(3):198–242.
51. Lee JH. The effects of music on pain: a meta-analysis. J Music Ther. 2016;53(4):430–77.
52. Mallik A, Chanda ML, Levitin DJ. Anhedonia to music and mu-opioids: evidence from the administration of naltrexone. Sci Rep. 2017;7:41952.

53. Lunde SJ, Vuust P, Garza-Villarreal EA, Kirsch I, Møller A, Vase L. Music-induced analgesia in healthy participants is associated with expected pain levels but not opioid or dopamine-dependent mechanisms. Front Pain Res. 2022;3:734999.
54. Rodgers-Melnick SN, Gam K, Debanne S, Little JA. Music use in adult patients with sickle cell disease: a pilot survey study. Music Ther Perspect. 2021;39(1):34–41.
55. Powers JM, Ioachim G, Stroman PW. Music to my senses: functional magnetic resonance imaging evidence of music analgesia across connectivity networks spanning the brain and brainstem. Front Pain Res. 2022;3:878258.
56. Ettenberger M, Maya R, Salgado-Vasco A, Monsalve-Duarte S, Betancourt-Zapata W, Suarez-Cañon N, et al. The effect of music therapy on perceived pain, mental health, vital signs, and medication usage of burn patients hospitalized in the intensive care unit: a randomized controlled feasibility study protocol. Front Psych. 2021;12:714209.
57. Scholz J. Mechanisms of chronic pain. Mol Pain. 2014;10(Suppl 1):O15. https://doi.org/10.1186/1744-8069-10-S1-O15.
58. Alshelh Z, Marciszewski KK, Akhter R, Di Pietro F, Mills EP, Vickers ER, Peck CC, Murray GM, Henderson LA. Disruption of default mode network dynamics in acute and chronic pain states. Neuroimage Clin. 2017;17:222–31.
59. Garza-Villarreal EA, Pando V, Vuust P, Parsons C. Music-induced analgesia in chronic pain conditions: a systematic review and meta-analysis. Pain Physician. 2017;20(7):597–610.
60. Howlin C, Rooney B. The cognitive mechanisms in music listening interventions for pain: a scoping review. J Music Ther. 2020;57(2):127–67.
61. Sena Moore K, Hanson-Abromeit D. Theory-guided therapeutic function of music to facilitate emotion regulation development in preschool-aged children. Front Hum Neurosci. 2015;9:572.
62. American Psychiatric Association. Science Watch: The two faces of oxytocin. Feb 2008 [cited 2023 Jan 11]. Available from https://www.apa.org/monitor/feb08/oxytocin.
63. Khalfa S, Bella SD, Roy M, Peretz I, Lupien SJ. Effects of relaxing music on salivary cortisol level after psychological stress. Ann N Y Acad Sci. 2003;999:374–6.
64. McKinney CH, Antoni MH, Kumar M, Tims FC, McCabe PM. Effects of guided imagery and music (GIM) therapy on mood and cortisol in healthy adults. Health Psychol. 1997;16(4):390–400.
65. Nilsson U. Soothing music can increase oxytocin levels during bed rest after open-heart surgery: a randomised control trial. J Clin Nurs. 2009;18(15):2153–61.
66. Ille R, Wabnegger A, Schwingenschuh P, Katschnig-Winter P, Kögl-Wallner M, Wenzel K, et al. Intact emotion recognition and experience but dysfunctional emotion regulation in idiopathic Parkinson's disease. J Neurol Sci. 2016;361:72–8.
67. Magee WL, Clark I, Tamplin J, Bradt J. Music interventions for acquired brain injury. Cochrane Database Syst Rev. 2017;(1):CD006787 [cited 2022 Sep 12]. Available from: https://www.cochranelibrary.com/cdsr/doi/10.1002/14651858.CD006787.pub3/full.
68. Ghazy AA, Soliman OA, Elbahnasi AI, Alawy AY, Mansour AM, Gowayed MA. Role of oxytocin in different neuropsychiatric, neurodegenerative, and neurodevelopmental disorders. In: Pedersen SHF, editor. Reviews of physiology, biochemistry and pharmacology, vol. 186. Cham: Springer; 2022. p. 95–134.
69. Blood AJ, Zatorre RJ. Intensely pleasurable responses to music correlate with activity in brain regions implicated in reward and emotion. Proc Natl Acad Sci. 2001;98(20):11818–23.
70. Tsoi KKF, Chan JYC, Ng YM, Lee MMY, Kwok TCY, Wong SYS. Receptive music therapy is more effective than interactive music therapy to relieve behavioral and psychological symptoms of dementia: a systematic review and meta-analysis. J Am Med Dir Assoc. 2018;19(7):568–576.e3.
71. van der Steen JT, Smaling HJ, van der Wouden JC, Bruinsma MS, Scholten RJ, Vink AC. Music-based therapeutic interventions for people with dementia. Cochrane Database Syst Rev. 2018;7:CD003477.
72. Legge AW. On the neural mechanisms of music therapy in mental health care: literature review and clinical implications. Music Ther Perspect. 2015;33(2):128–41.

73. Kang K, Orlandi S, Lorenzen N, Chau T, Thaut M. Does music induce interbrain synchronization between a non-speaking youth with cerebral palsy (CP), a parent, and a neurologic music therapist? A brief report. Dev Neurorehabil. 2022;25(6):426–32.
74. Samadani A, Kim S, Moon J, Kang K, Chau T. Neurophysiological synchrony between children with severe physical disabilities and their parents during music therapy. Front Neurosci. 2021;15:531915.
75. Trehub SE, Becker J, Morley I. Cross-cultural perspectives on music and musicality. Philos Trans R Soc Lond Ser B Biol Sci. 2015;370(1664):20140096.
76. Robb SL, Burns DS, Carpenter JS. Reporting guidelines for music-based interventions. J Health Psychol. 2011;16(2):342–52. Epub 2010 Aug 13. Erratum in: J Health Psychol. 2011 Mar;16(2):393.
77. Hakvoort L, Tönjes J. Music-mechanisms at the core of music therapy: towards a format for a description of music therapy micro-interventions. Nord J Music Ther. 2023;32(1):67–91.
78. Edwards E, St Hillaire-Clarke C, Frankowski DW, Finkelstein R, Cheever T, Chen WG, et al. NIH music-based intervention toolkit: music-based interventions for brain disorders of aging. Neurology. 2023;100(18):868–78.

Music for Stroke Rehabilitation

Anna Palumbo, Soo Ji Kim, and Preeti Raghavan

Introduction

Stroke is a leading cause of disability, and often results in multiple impairments that impact motor function, cognition, language, activities of daily living, and emotional well-being [1–6]. Music therapy and music-based interventions address each of these domains, sometimes with a single intervention [7]. Music listening has shown benefits for mood, cognition, language, and visual neglect [8–10]. Rhythmic auditory stimulation (RAS®) utilizes musical cues to enhance gait performance [11]. Playing musical instruments may lead to improvements in upper limb motor recovery, spatial neglect, and emotional and social well-being [12, 13]. Among individuals with poststroke aphasia, Melodic Intonation Therapy (MIT) enhances functional communication, and group singing supports emotional well-being and social interactions [11, 14]. Finally, technology-assisted music interventions encourage functional task repetition and awareness of arm position through music and sound [15, 16].

A. Palumbo (✉)
Rehabilitation Sciences Program, Steinhardt School of Education, Culture, and Human Development, New York University, New York, NY, USA
e-mail: Ap89@nyu.edu

S. J. Kim
Music Therapy Education, Graduate School of Education, Ewha Womans University, Seoul, South Korea
e-mail: specare@ewha.ac.kr

P. Raghavan
Departments of Physical Medicine and Rehabilitation and Neurology, Johns Hopkins University School of Medicine, Baltimore, MD, USA
e-mail: praghavan@jhmi.edu

© The Author(s), under exclusive license to Springer Nature Switzerland AG 2023
K. Devlin et al. (eds.), *Music Therapy and Music-Based Interventions in Neurology*, Current Clinical Neurology,
https://doi.org/10.1007/978-3-031-47092-9_3

Self-Location

Anna Palumbo, MA, MT-BC, LCAT, is a board-certified music therapist, licensed creative arts therapist, and doctoral candidate in rehabilitation science. Her practice has included music psychotherapy and the use of music in neurorehabilitation, dementia care, acute mental illness, early childhood development, and work with neurodiverse populations. Her research investigates the physical, emotional, and social processes during clinical improvisation in music therapy for stroke rehabilitation.

Soo Ji Kim, PhD, MT-BC, KCMT, is a professor and director of music therapy. Dr. Kim specializes in music therapy for neurorehabilitation and cognitive aging. She integrates neuroscience, biopsychosocial care, and medical music therapy in her research, focusing on the connection between music and the brain. Her work emphasizes the importance of evidence-based practices in music therapy. Dr. Kim has extensive clinical experience with diverse populations, including adults with neurological disorders, older adults in community settings, and patients receiving dementia care and their families.

Preeti Raghavan, MBBS, is a physician-scientist in physical medicine and rehabilitation specializing in stroke recovery and rehabilitation. She is board certified in physical medicine and rehabilitation and brain injury medicine. She completed a research fellowship in motor control and learning and has been an NIH-funded researcher for the last 20 years. Dr. Raghavan has developed several innovative approaches to rehabilitation from stroke and other brain injuries including an enriched collaborative approach to rehabilitation for stroke using music and movement, called music upper limb therapy-integrated (MULT-I). Her research interests include the recovery of arm and hand function after brain injury, interactions between emotional regulation and recovery, treatment of muscle stiffness, and improvements to stroke systems of care to mitigate disability.

Definitions

Stroke

A stroke refers to neurological dysfunction caused by a focal cerebral, spinal, or retinal infarction or hemorrhage [17]. Both infarction and hemorrhage reduce cerebral blood flow (CBF) resulting in brain cell death. The stages of recovery after stroke have been defined as acute, subacute and chronic—corresponding to the first 7 days, day 7 through the sixth month, and beyond the sixth month, respectively—based upon biological recovery, including cell death, inflammation and scarring, neural plasticity and functional impairment [18].

Neurorehabilitation

Neurorehabilitation involves the comprehensive assessment of impairment in body structures and function, activity limitations, and participation restrictions by an interprofessional team, which may include a rehabilitation physician, a nurse, physical therapist, occupational therapist, speech/language therapist, psychologist, social worker, and others (e.g. dietician, orthotist), focusing on early mobilization to prevent the complications of immobility, and to restore movement, activity, and participation over the long term [19]. While some rehabilitation may begin during the acute hospital stay, most rehabilitation occurs after acute hospital discharge, either to home or to a rehabilitation facility [19]. A major challenge in neurorehabilitation is to provide adequate therapy to address the physical, emotional, and cognitive effects of stroke. Music therapy and music-based interventions address each of these areas of neurorehabilitation by facilitating repetitive movements through auditory-motor coupling [20–22], enhancing mood and cognitive processing [9, 23, 24], and reinforcing social bonds through joint music-making [25–28].

Literature Review

Music Listening

Listening to preferred music has been linked to improved cognition and language skills, reduced depression and visual neglect, and neuroplastic changes in frontal, motor, language, and emotion-related brain areas following a stroke [8–10]. In a randomized controlled trial (RCT), listening to preferred music during subacute stroke rehabilitation improved cognition—including verbal memory and focused attention—and reduced depressed affect compared to listening to audio books or to no auditory stimuli [9]. The improvements correlated with increased grey matter volume in frontal and limbic areas that play a role in working memory and reward and emotion-related responses, respectively [10]. Follow up studies have demonstrated that listening to vocal music may improve verbal memory and language skills more than instrumental music or audiobooks, especially among individuals with poststroke aphasia [8, 29]. However, listening to music—either instrumental or vocal—is associated with increased structural connectivity in white matter tracts in language and motor areas [30]. Listening to preferred music can also reduce visual neglect compared to non-preferred music [31, 32], white noise [31], or silence [32]. The reduction in visual neglect during listening to preferred music was greater when music was presented with spatial cues, such as when the sound stimuli moved continuously from the right to the left ear [33, 34]. Voxel based lesion symptom mapping linked responsiveness to auditory spatial cues with the integrity of brain areas involved with sound localization, visual attention, and interhemispheric communication [33].

Moving to Musical Cues

Rhythmic auditory stimulation (RAS®) is a neurologic music therapy (NMT®) technique used in physical rehabilitation to improve motor control, especially gait [35]. It modulates motor neuron activity and entrains neural activation patterns through regularly timed rhythmic cueing [11]. Several studies have demonstrated the effectiveness of RAS® for improving functional gait performance during both overground and treadmill walking [36–39], and upper limb motor function, compared to conventional rehabilitation [40]. External rhythmic cues and rhythmic auditory cues have also been used outside of RAS® [36]. Comparison across these studies demonstrates that combining rhythmic cues with music significantly improves gait velocity, compared to rhythmic cueing alone, and that the improvement is greater when the intervention is administered by a music therapist [36]. However, heterogeneity in effect sizes across studies reflects the need for standardized definitions of interventions using rhythmic cues to allow generalizability of findings [37]. Perception of the beat presented through music remains the key in RAS®. Digital technologies (see Chap. 14) are increasingly being incorporated to provide RAS®-driven gait training, where wearable sensors provide real-time movement feedback to individualize the rhythmic beat and automate progressive gait training for patients at home and in the community poststroke [41, 42].

Playing Musical Instruments

Playing musical instruments has been utilized in stroke rehabilitation to improve motor functioning, cognition, social interaction, and emotional well-being, as well as to reduce spatial neglect. Several music-playing interventions have been developed, including Music Therapeutic Instrumental Performance (TIMP®), Music Supported Therapy (MST), Music Upper Limb Therapy (MULT-I), and approaches to reduce spatial neglect (e.g., Musical Neglect Training®).

Therapeutic Instrumental Music Performance (TIMP®)

TIMP® is a NMT® technique in which percussion instruments are played to target specific patterns of limb movement and range of motion, while leveraging musical structure to cue movement patterns [11, 43]. Similar to RAS®, TIMP® utilizes a strong and precise rhythmic pulse to cue movement based on principles of auditory-motor coupling [11, 43]. In a recent pilot study, TIMP® was provided in one-on-one sessions and led to improved motor function, although this gain was not tested for significance due to small sample size [44]. Participants described the rhythmic structure as helpful for coordinating their movement and feelings of motivation and challenge during the intervention [44]. An RCT that compared TIMP® alone to TIMP® combined with metronome-cued motor imagery (TIMP®-cMI) or non-cued motor imagery (TIMP®-MI) in stroke survivors showed reduced motor impairment in all groups, with the TIMP® group exhibiting greater improvement than the TIMP®-cMI group [45]. The TIMP-cMI group showed positive changes in affect, while the TIMP®-MI group demonstrated increased mental flexibility, suggesting that combining TIMP® with motor imagery may interfere with gains in motor recovery but provide benefits for cognition and affect poststroke [46].

Music Supported Therapy (MST)
MST involves playing melodies on either a piano or using drum pads tuned to a musical scale to develop fine and gross motor movement, respectively [22, 47]. Melodies familiar to participants are selected, and the difficulty of the music-playing task is adjusted to meet functional rehabilitation goals [22, 47]. An early study demonstrated that MST improved fine upper limb motor function more than dose matched constraint-induced therapy when both were offered in addition to conventional therapy during inpatient stroke rehabilitation [48]. In subsequent trials, MST improved upper limb motor function to the same level as additional dose-matched exercise and task-based interventions [12, 49]; however, motor recovery in the MST group improved more for those with greater reward response when moving to music, as measured at baseline using the Barcelona Music Reward Questionnaire [12]. From pre- to post-intervention, MST is associated with improved cognition, mood, and quality of life, with significant improvement in verbal learning compared to healthy controls [12, 50, 51]. Neurologically, MST has been linked to increased activation and reorganization of the affected sensorimotor cortex, reduced compensatory activation of the unaffected hemisphere, and improved functional connectivity between motor and auditory pathways in chronic stroke survivors [50, 52, 53].

Music Upper Limb Therapy-Integrated (MULT-I)
MULT-I creates an enriched environment for stroke rehabilitation that targets physical, emotional, and social aspects of stroke recovery using collaborative group music-making [13]. During MULT-I, stroke survivors play instruments together in a group co-led by a music therapist and occupational therapist, following the Nordoff-Robbins approach to music therapy [28]. This approach supports the creative process of music-making through music improvisation [28, 54]. In an open label study, MULT-I led to improved motor recovery, tactile sensitivity, and emotional well-being that was maintained at 1-year follow up [13]. In a follow-up RCT, MULT-I led to reduced levels of depression and increased serum levels of brain derived neurotrophic factor, which is associated with improved stroke recovery outcomes, compared to a home exercise program without music and social enrichment [55]. Participants described the process of group music-making as generating peer support and a sense of belonging, motivating them to take on challenges without fear of failure, and increasing emotional awareness, expression, and enjoyment [28]. Taken together, these findings suggest that MULT-I effectively integrates social and emotional aspects of stroke rehabilitation with motor recovery [56].

Music Playing to Reduce Spatial Neglect
Playing pitches in a scale or a familiar song has been shown to reduce spatial neglect following a stroke. Individuals with left spatial neglect show improved exploration of the left side when playing a descending scale on piano, as compared to playing

piano in silence or with randomized tones [57]. In two similar case studies using tone bars using Musical Neglect Training (MNT®), playing a scale or familiar song resulted in reduced visual neglect, even when not playing the instrument [58, 59]. These findings suggest that playing musical instruments increases attention to the neglected visual field.

Singing

Melodic Intonation Therapy (MIT)

MIT (see Chap. 6) is a systematic structured method that teaches individuals with non-fluent aphasia to produce intoned (sung) patterns while tapping each syllable with the left hand [60]. This method draws on the preserved ability to sing in some individuals with non-fluent aphasia who otherwise experience impairment in expressive language [61]. Intoned patterns are designed to exaggerate the melodic components of speech, and progress from 1 to 2 syllables to longer sentences that express daily needs [11]. Rhythmic tapping with the left hand is designed to support priming of orofacial and articulatory movements by providing temporal pacing and activation of non-lesioned brain areas [62]. A systematic review of four RCTs demonstrated the positive effects of MIT on functional communication and repetition [63]. The neural mechanisms of these effects continue to be investigated [64]. In a recent open label study, MIT increased activity in right hemisphere regions including the inferior frontal gyrus, which correlated with improvements in speech fluency among stroke survivors with chronic, moderate-to-severe non-fluent aphasia with large left hemisphere lesions [65]. However, other studies have identified increased left hemisphere or bilateral brain activity following MIT, suggesting neural mechanisms for recovery may vary in relation to lesion size and other individual factors [64].

Group Singing

Group singing has been linked to improved social cohesion, suggesting that it can play a role in providing social support following stroke [66]. Participation in social activities, including group singing, is associated with improved functional communication poststroke [67]. Participants in group singing report improved mood, confidence in verbal communication, and peer support, addressing social isolation and other common difficulties following a stroke [14, 68, 69]. However, more studies are needed to understand the mechanisms driving social and functional improvements associated with group singing.

Technology-Assisted Interventions

MusicGlove

The MusicGlove facilitates practice of gripping-like movements and thumb-finger opposition in a music-based video game. In an initial study, the MusicGlove led to improvements in hand function that were sustained at 1-month follow-up compared to conventional hand therapy among survivors of stroke with mild-to-moderate impairment in the chronic stage [15]. MusicGlove scores correlated strongly with gross motor function, and participants described MusicGlove as more motivating than conventional hand therapy [15]. In a subsequent home-based clinical trial, the MusicGlove led to similar gains in hand function as conventional therapy, but the MusicGlove led to significantly greater reports of functional use and quality of movement [70]. Participants in MusicGlove significantly increased the number of grips completed after the first week, suggesting good treatment adherence [70]. These data suggest that technology that incorporates music can lead to greater motivation and repetitions that are thought to be essential for recovery.

Sonification of Arm Movements

Musical sonification therapy was designed to provide real-time musical feedback about the position of the arm in space to motivate gross arm movements early after stroke [71]. Two initial studies compared music sonification therapy to sham sonification training, in which the same movements were completed without sound, and showed similar effects in motor recovery for both groups, although musical sonification led to significantly reduced joint pain and increased smoothness of movement [71, 72]. A recent RCT found that sonification treatment led to increased gains in motor recovery compared to dose-matched standard therapy, when both were provided in addition to conventional rehabilitation [16]. Together, these findings suggest that combining musical feedback with conventional rehabilitation may improve upper extremity rehabilitation poststroke.

Case Example(s)

Case 1: RAS® Gait Training

K.H., a 64-year-old male diagnosed with right hemiplegia from intracerebral hemorrhage, sought music therapy for gait training at a hospital in Seoul, Korea. He had weakness on the right side of his trunk and an unstable gait with leg circumduction. Additionally, he experienced global aphasia, which hindered communication and therapy. Despite attending physical and occupational therapy, K.H. and his daughter, his primary caregiver, expressed frustration and depression due to his limited understanding of verbal instructions. K.H. enjoyed hymns and other religious music as he had been a devout Christian for most of his life.

After a music therapy intake session with K.H. and his daughter, the goal was set to improve K.H.'s gait stability and speed using music therapy. An RAS® protocol was designed over 12 30-min sessions. The first session included a lab-based gait assessment, and subsequent RAS® sessions were conducted in collaboration with K.H.'s physical therapist. Each session involved assessing cadence, pre-gait training, practicing gait with RAS®, and practicing gait without RAS®. K.H.'s initial cadence was 66 steps/min.

Fig. 3.1 This is an example of the music used during RAS® sessions with K.H. (**a**). The original hymn [What a Fellowship, What a Joy Divine in A] was arranged to be a cue for walking (**b**)

K.H. was initially unresponsive to verbal directions and showed minimal facial expressions during music therapy sessions. Live musical accompaniment was used to provide movement cues. Each session involved 3–5 min of pre-gait training followed by 15–20 min of walking with RAS®. A cane was utilized for stability, and K.H. demonstrated synchronization between his gait and the musical cues. The therapist selected RAS® music, including hymns, based on K.H.'s preferences (Fig. 3.1).

Over the course of 12 sessions, K.H. also exhibited emotional changes such as becoming tearful in response to specific music. His daughter noticed positive behavioral changes, including responsiveness to the therapist's cues and expression of his musical preferences. As the sessions progressed, K.H.'s cadence improved, reaching 85 steps/min on the final assessment. Furthermore, he achieved balanced weight distribution while walking.

Case 2: MULT-I

Sharlene[1] was a 58-year-old female who experienced a left middle cerebral artery stroke causing unilateral hemiparesis and non-fluent aphasia, limiting both her movement and expressive communication. Sharlene participated in a MULT-I group

---

[1] Pseudonym.

with 4 other survivors of stroke 5 months after her stroke. At the beginning of the intervention, Sharlene had a patient health questionnaire-9 (PHQ-9) score of 17, indicating moderately severe depression. She arrived slouched in her wheelchair, appeared withdrawn, and initially declined to participate in music-making or accept support from the occupational therapist. The music therapist encouraged Sharlene's development by incorporating her verbal expressions into improvisations and creating space for her to be heard in the music. Gradually, Sharlene accepted support from the occupational therapist to play instruments, made an effort to play instruments independently, and began singing along with improvised songs, thus working on her verbal expression in parallel with her physical movement. Sharlene eventually became a leader for the group, encouraging other group members to sing, making specific song requests, and directing the finale of group improvisations. Other group members supported Sharlene by enthusiastically following her direction and celebrating her achievements. In the final session, another participant reflected on how Sharlene had "really come out of [her] shell and was doing well," and that, "it made everybody happy to see" these changes. In her final session, Sharlene worked to find the words to express gratitude for her time in the group and shared, "I want to say… thank you for give us."

Over the course of 9 sessions, Sharlene's depression levels reduced to the border of mild to moderate depression with a PHQ-9 score of 10, a 7-point reduction in her PHQ-9 score in 6 weeks. She additionally rated herself as having higher self-efficacy and quality of life, as well as improved communication, activities of daily living, participation in her societal roles, physical strength, and recovery from her stroke on the stroke impact scale. Although her level of disability on the modified Rankin Scale remained unchanged, her emotional and social wellbeing and ability to engage meaningfully in daily life activities had greatly improved.

Conclusion and Future Considerations

Music interventions for stroke rehabilitation have been implemented using a variety of methods, including music listening, instrument playing, responding to rhythmic cues, and singing. Each of these methods offers potential to enhance rehabilitation outcomes, with different approaches emphasizing improvement in motor recovery, expressive language, spatial awareness, and emotional and social well-being. Several interventions, even passive music listening, have demonstrated potential to enhance multiple domains of stroke rehabilitation, suggesting the potential for music to play an important role in effective interdisciplinary treatment. There has been growing collaboration between the medical, neuroscience, rehabilitation and music therapy fields leading to more efficient and effective applications of music therapy in stroke recovery. Future research will further develop the role of music in this area, elucidating underlying mechanisms, clarifying optimal music intervention dosage and addressing the use of technology and home-based music therapy.

References

1. WHO. Global health estimates 2020: disease burden by cause, age, sex, by country and by region, 2000-2019. World Health Organization; 2020. Available from: https://www.who.int/data/gho/data/themes/mortality-and-global-health-estimates/global-health-estimates-leading-causes-of-dalys.
2. Johnson CO, Nguyen M, Roth GA, Nichols E, Alam T, Abate D, et al. Global, regional, and national burden of stroke, 1990–2016: a systematic analysis for the Global Burden of Disease Study 2016. Lancet Neurol. 2019;18(5):439–58.
3. Northcott S, Moss B, Harrison K, Hilari K. A systematic review of the impact of stroke on social support and social networks: associated factors and patterns of change. Clin Rehabil. 2016;30(8):811–31.
4. Schnitzler A, Jourdan C, Josseran L, Azouvi P, Jacob L, Genêt F. Participation in work and leisure activities after stroke: a national study. Ann Phys Rehabil Med. 2019;62(5):351–5.
5. Ayerbe L, Ayis S, Wolfe CD, Rudd AG. Natural history, predictors and outcomes of depression after stroke: systematic review and meta-analysis. Br J Psychiatry. 2013;202(1):14–21.
6. Ayerbe L, Ayis SA, Crichton S, Wolfe CD, Rudd AG. Natural history, predictors and associated outcomes of anxiety up to 10 years after stroke: the South London Stroke Register. Age Ageing. 2014;43(4):542–7.
7. Sihvonen AJ, Särkämö T, Leo V, Tervaniemi M, Altenmüller E, Soinila S. Music-based interventions in neurological rehabilitation. Lancet Neurol. 2017;16(8):648–60.
8. Sihvonen AJ, Leo V, Ripollés P, Lehtovaara T, Ylönen A, Rajanaro P, et al. Vocal music enhances memory and language recovery after stroke: pooled results from two RCTs. Ann Clin Transl Neurol. 2020;7(11):2272–87.
9. Sarkamo T, Tervaniemi M, Laitinen S, Forsblom A, Soinila S, Mikkonen M, et al. Music listening enhances cognitive recovery and mood after middle cerebral artery stroke. Brain. 2008;131(Pt 3):866–76.
10. Sarkamo T, Ripolles P, Vepsalainen H, Autti T, Silvennoinen HM, Salli E, et al. Structural changes induced by daily music listening in the recovering brain after middle cerebral artery stroke: a voxel-based morphometry study. Front Hum Neurosci. 2014;8:245.
11. Thaut MH, Hoemberg V. Handbook of neurologic music therapy. Oxford: Oxford University Press; 2014.
12. Grau-Sanchez J, Duarte E, Ramos-Escobar N, Sierpowska J, Rueda N, Redon S, et al. Music-supported therapy in the rehabilitation of subacute stroke patients: a randomized controlled trial. Ann N Y Acad Sci. 2018;1423:318–28.
13. Raghavan P, Geller D, Guerrero N, Aluru V, Eimicke JP, Teresi JA, et al. Music upper limb therapy-integrated: an enriched collaborative approach for stroke rehabilitation. Front Hum Neurosci. 2016;10:498.
14. Fogg-Rogers L, Buetow S, Talmage A, McCann CM, Leao SH, Tippett L, et al. Choral singing therapy following stroke or Parkinson's disease: an exploration of participants' experiences. Disabil Rehabil. 2016;38(10):952–62.
15. Friedman N, Chan V, Reinkensmeyer AN, Beroukhim A, Zambrano GJ, Bachman M, et al. Retraining and assessing hand movement after stroke using the MusicGlove: comparison with conventional hand therapy and isometric grip training. J Neuroeng Rehabil. 2014;11(1):76.
16. Raglio A, Panigazzi M, Colombo R, Tramontano M, Iosa M, Mastrogiacomo S, et al. Hand rehabilitation with sonification techniques in the subacute stage of stroke. Sci Rep. 2021;11(1):7237.
17. Sacco RL, Kasner SE, Broderick JP, Caplan LR, Connors JJ, Culebras A, et al. An updated definition of stroke for the 21st century: a statement for healthcare professionals from the American Heart Association/American Stroke Association. Stroke. 2013;44(7):2064–89.
18. Bernhardt J, Hayward KS, Kwakkel G, Ward NS, Wolf SL, Borschmann K, et al. Agreed definitions and a shared vision for new standards in stroke recovery research: the stroke recovery and rehabilitation roundtable taskforce. Int J Stroke. 2017;12(5):444–50.

19. Winstein CJ, Stein J, Arena R, Bates B, Cherney LR, Cramer SC, et al. Guidelines for adult stroke rehabilitation and recovery: a guideline for healthcare professionals from the American Heart Association/American Stroke Association. Stroke. 2016;47(6):e98–e169.
20. Roerdink M, Lamoth CJ, van Kordelaar J, Elich P, Konijnenbelt M, Kwakkel G, et al. Rhythm perturbations in acoustically paced treadmill walking after stroke. Neurorehabil Neural Repair. 2009;23(7):668–78.
21. Rojo N, Amengual J, Juncadella M, Rubio F, Camara E, Marco-Pallares J, et al. Music-supported therapy induces plasticity in the sensorimotor cortex in chronic stroke: a single-case study using multimodal imaging (fMRI-TMS). Brain Inj. 2011;25(7–8):787–93.
22. Rodriguez-Fornells A, Rojo N, Amengual JL, Ripolles P, Altenmuller E, Munte TF. The involvement of audio-motor coupling in the music-supported therapy applied to stroke patients. Ann N Y Acad Sci. 2012;1252:282–93.
23. Malcolm MP, Massie C, Thaut M. Rhythmic auditory-motor entrainment improves hemiparetic arm kinematics during reaching movements: a pilot study. Top Stroke Rehabil. 2009;16(1):69–79.
24. Särkämö T, Soto D. Music listening after stroke: beneficial effects and potential neural mechanisms. Ann N Y Acad Sci. 2012;1252:266–81.
25. Hove MJ, Risen JL. It's all in the timing: interpersonal synchrony increases affiliation. Soc Cogn. 2009;27(6):949–60.
26. Miles LK, Nind LK, Macrae CN. The rhythm of rapport: interpersonal synchrony and social perception. J Exp Soc Psychol. 2009;45(3):585–9.
27. Cirelli LK, Einarson KM, Trainor LJ. Interpersonal synchrony increases prosocial behavior in infants. Dev Sci. 2014;17(6):1003–11.
28. Guerrero N, Turry A, Geller D, Raghavan P. From historic to contemporary: Nordoff-Robbins music therapy in collaborative interdisciplinary rehabilitation. Music Ther Perspect. 2014;32(1):38–46.
29. Sihvonen AJ, Ripollés P, Leo V, Saunavaara J, Parkkola R, Rodríguez-Fornells A, et al. Vocal music listening enhances post-stroke language network reorganization. eNeuro. 2021;8(4):ENEURO.0158-21.2021.
30. Sihvonen AJ, Soinila S, Särkämö T. Post-stroke enriched auditory environment induces structural connectome plasticity: secondary analysis from a randomized controlled trial. Brain Imaging Behav. 2022;16(4):1813–22.
31. Chen M-C, Tsai P-L, Huang Y-T, Lin K-C. Pleasant music improves visual attention in patients with unilateral neglect after stroke. Brain Inj. 2013;27(1):75–82.
32. Soto D, Funes MJ, Guzman-Garcia A, Warbrick T, Rotshtein P, Humphreys GW. Pleasant music overcomes the loss of awareness in patients with visual neglect. Proc Natl Acad Sci U S A. 2009;106(14):6011–6.
33. Kaufmann BC, Cazzoli D, Bartolomeo P, Frey J, Pflugshaupt T, Knobel SEJ, et al. Auditory spatial cueing reduces neglect after right-hemispheric stroke: a proof of concept study. Cortex. 2022;148:152–67.
34. Schenke N, Franke R, Puschmann S, Turgut N, Kastrup A, Thiel CM, et al. Can auditory cues improve visuo-spatial neglect? Results of two pilot studies. Neuropsychol Rehabil. 2021;31(5):710–30.
35. Thaut MH. The discovery of human auditory-motor entrainment and its role in the development of neurologic music therapy. Prog Brain Res. 2015;217:253–66.
36. Magee WL, Clark I, Tamplin J, Bradt J. Music interventions for acquired brain injury. Cochrane Database Syst Rev. 2017;1:CD006787.
37. Yoo GE, Kim SJ. Rhythmic auditory cueing in motor rehabilitation for stroke patients: systematic review and meta-analysis. J Music Ther. 2016;53(2):149–77.
38. Hollands KL, Pelton TA, Tyson SF, Hollands MA, van Vliet PM. Interventions for coordination of walking following stroke: systematic review. Gait Posture. 2012;35(3):349–59.
39. Thaut MH, Leins AK, Rice RR, Argstatter H, Kenyon GP, McIntosh GC, et al. Rhythmic auditory stimulation improves gait more than NDT/Bo bath training in near-ambulatory

patients early poststroke: a single-blind, randomized trial. Neurorehabil Neural Repair. 2007;21(5):455–9.
40. Tian R, Zhang B, Zhu Y. Rhythmic auditory stimulation as an adjuvant therapy improved post-stroke motor functions of the upper extremity: a randomized controlled pilot study. Front Neurosci. 2020;14:649.
41. Harris B, Awad L. Automating a progressive and individualized rhythm-based walking training program after stroke: feasibility of a music-based digital therapeutic. Arch Phys Med Rehabil. 2020;101(11):e30.
42. Hutchinson K, Sloutsky R, Collimore A, Adams B, Harris B, Ellis TD, et al. A music-based digital therapeutic: proof-of-concept automation of a progressive and individualized rhythm-based walking training program after stroke. Neurorehabil Neural Repair. 2020;34(11):986–96.
43. Thaut MH. Rhythm, music, and the brain: scientific foundations and clinical applications. New York, NY: Routledge; 2008.
44. Street AJ, Magee WL, Bateman A, Parker M, Odell-Miller H, Fachner J. Home-based neurologic music therapy for arm hemiparesis following stroke: results from a pilot, feasibility randomized controlled trial. Clin Rehabil. 2018;32(1):18–28.
45. Haire CM, Tremblay L, Vuong V, Patterson KK, Chen JL, Burdette JH, et al. Therapeutic instrumental music training and motor imagery in post-stroke upper-extremity rehabilitation: a randomized-controlled pilot study. Arch Rehabil Res Clin Transl. 2021;3(4):100162.
46. Haire CM, Vuong V, Tremblay L, Patterson KK, Chen JL, Thaut MH. Effects of therapeutic instrumental music performance and motor imagery on chronic post-stroke cognition and affect: a randomized controlled trial. NeuroRehabilitation. 2021;48(2):195–208.
47. Schneider S, Schonle PW, Altenmuller E, Munte TF. Using musical instruments to improve motor skill recovery following a stroke. J Neurol. 2007;254(10):1339–46.
48. Schneider S, Müünte T, Rodriguez-Fornells A, Sailer M, Altenmüüller E. Music-supported training is more efficient than functional motor training for recovery of fine motor skills in stroke patients. Music Percept. 2010;27(4):271–80.
49. Fujioka T, Dawson DR, Wright R, Honjo K, Chen JL, Chen JJ, et al. The effects of music-supported therapy on motor, cognitive, and psychosocial functions in chronic stroke. Ann N Y Acad Sci. 2018;1423:264–74.
50. Ripolles P, Rojo N, Grau-Sanchez J, Amengual JL, Camara E, Marco-Pallares J, et al. Music supported therapy promotes motor plasticity in individuals with chronic stroke. Brain Imaging Behav. 2016;10(4):1289–307.
51. Van Vugt FT, Ritter J, Rollnik JD, Altenmuller E. Music-supported motor training after stroke reveals no superiority of synchronization in group therapy. Front Hum Neurosci. 2014;8:315.
52. Amengual JL, Rojo N, Veciana de Las Heras M, Marco-Pallares J, Grau-Sanchez J, Schneider S, et al. Sensorimotor plasticity after music-supported therapy in chronic stroke patients revealed by transcranial magnetic stimulation. PLoS One. 2013;8(4):e61883.
53. Grau-Sanchez J, Amengual JL, Rojo N, Veciana de Las Heras M, Montero J, Rubio F, et al. Plasticity in the sensorimotor cortex induced by music-supported therapy in stroke patients: a TMS study. Front Hum Neurosci. 2013;7:494.
54. Nordoff P, Robbins C. Creative music therapy: a guide to fostering clinical musicianship. 2nd ed. revised. Gilsum, NH: Barcelona Publishers; 2007.
55. Palumbo A, Aluru V, Battaglia J, Geller D, Turry A, Ross M, et al. Music upper limb therapy-integrated (MULT-I) provides a feasible enriched environment and reduces post stroke depression: a pilot randomized controlled trial. Am J Phys Med Rehabil. 2022;101(10):937–46.
56. Corbett D, Jeffers M, Nguemeni C, Gomez-Smith M, Livingston-Thomas J. Lost in translation: rethinking approaches to stroke recovery. Prog Brain Res. 2015;218:413–34.
57. Bernardi NF, Cioffi MC, Ronchi R, Maravita A, Bricolo E, Zigiotto L, et al. Improving left spatial neglect through music scale playing. J Neuropsychol. 2017;11(1):135–58.
58. Kang K, Thaut MH. Musical neglect training for chronic persistent unilateral visual neglect post-stroke. Front Neurol. 2019;10:474.

59. Bodak R, Malhotra P, Bernardi NF, Cocchini G, Stewart L. Reducing chronic visuo-spatial neglect following right hemisphere stroke through instrument playing. Front Hum Neurosci. 2014;8:413.
60. Albert ML, Sparks RW, Helm NA. Melodic intonation therapy for aphasia. Arch Neurol. 1973;29(2):130–1.
61. NIDCD. Aphasia: National Institute on Deafness and Other Communication Disorders; 2015. Available from: https://www.nidcd.nih.gov/health/aphasia.
62. Curtis S, Nicholas ML, Pittmann R, Zipse L. Tap your hand if you feel the beat: differential effects of tapping in melodic intonation therapy. Aphasiology. 2020;34(5):580–602.
63. Haro-Martínez A, Pérez-Araujo CM, Sanchez-Caro JM, Fuentes B, Díez-Tejedor E. Melodic intonation therapy for post-stroke non-fluent aphasia: systematic review and meta-analysis. Front Neurol. 2021;12:700115.
64. García-Casares N, Barros-Cano A, García-Arnés JA. Melodic intonation therapy in post-stroke non-fluent aphasia and its effects on brain plasticity. J Clin Med. 2022;11(12):3503.
65. Marchina S, Norton A, Schlaug G. Effects of melodic intonation therapy in patients with chronic nonfluent aphasia. Ann N Y Acad Sci. 2023;1519(1):173–85.
66. Tarrant M, Lamont RA, Carter M, Dean SG, Spicer S, Sanders A, et al. Measurement of shared social identity in singing groups for people with aphasia. Front Psychol. 2021;12:669899.
67. Zumbansen A, Peretz I, Anglade C, Bilodeau J, Généreux S, Hubert M, et al. Effect of choir activity in the rehabilitation of aphasia: a blind, randomised, controlled pilot study. Aphasiology. 2017;31(8):879–900.
68. Tamplin J, Baker FA, Jones B, Way A, Lee S. 'Stroke a Chord': the effect of singing in a community choir on mood and social engagement for people living with aphasia following a stroke. NeuroRehabilitation. 2013;32(4):929–41.
69. Tarrant M, Carter M, Dean SG, Taylor R, Warren FC, Spencer A, et al. Singing for people with aphasia (SPA): results of a pilot feasibility randomised controlled trial of a group singing intervention investigating acceptability and feasibility. BMJ Open. 2021;11(1):e040544.
70. Zondervan DK, Friedman N, Chang E, Zhao X, Augsburger R, Reinkensmeyer DJ, et al. Home-based hand rehabilitation after chronic stroke: randomized, controlled single-blind trial comparing the MusicGlove with a conventional exercise program. J Rehabil Res Dev. 2016;53(4):457–72.
71. Scholz DS, Rohde S, Nikmaram N, Brückner H-P, Großbach M, Rollnik JD, et al. Sonification of arm movements in stroke rehabilitation—a novel approach in neurologic music therapy. Front Neurol. 2016;7:106.
72. Nikmaram N, Scholz DS, Grossbach M, Schmidt SB, Spogis J, Belardinelli P, et al. Musical sonification of arm movements in stroke rehabilitation yields limited benefits. Front Neurosci. 2019;13:1378.

Music for Traumatic Brain Injury and Impaired Consciousness

4

Jeanette Tamplin, Janeen Bower, and Sini-Tuuli Siponkoski

Introduction

Common disciplinary approaches in traumatic brain injury (TBI) rehabilitation include neuropsychology, occupational therapy, physiotherapy, and speech therapy. There is growing evidence for music-based approaches in neurorehabilitation and music therapy is now included in many multi-disciplinary rehabilitation programs. The combination of rehabilitation methods used depend often on the severity level of the injury and stage of recovery. Very early on, when the patient is still medically unstable, amnesic and/or experiencing a disorders of consciousness (DoC), it is of primary importance to establish the patient's physical and emotional safety through personalised medical care and environmental factors and to build a reliable and consistent means of communication and strengthen orientation. As the patient's own understanding of their situation emerges, and they are medically stabilized, the more active phase of rehabilitation can begin. Although neuroplasticity supports neural recovery best within months of the injury [1], psychological adjustment can take much longer. More holistic, multi-professional approaches incorporating music

---

J. Tamplin (✉)
Faculty of Fine Arts and Music, University of Melbourne, Melbourne, Australia

Austin Health, Melbourne, Australia
e-mail: jeanette.tamplin@unimelb.edu.au

J. Bower
Faculty of Fine Arts and Music, University of Melbourne, Melbourne, Australia

Royal Children's Hospital, Melbourne, Australia
e-mail: janeen.bower@rch.org.au

S.-T. Siponkoski
The University of Helsinki, Helsinki, Finland
e-mail: sini-tuuli.siponkoski@helsinki.fi

© The Author(s), under exclusive license to Springer Nature Switzerland AG 2023
K. Devlin et al. (eds.), *Music Therapy and Music-Based Interventions in Neurology*, Current Clinical Neurology,
https://doi.org/10.1007/978-3-031-47092-9_4

have been successfully used to improve outcomes following TBI even years after initial injury [2]. In this chapter, we will explore the therapeutic application of music in neurorehabilitation, with a specific focus on its applications in cognitive rehabilitation and regulating arousal and awareness for people with DoC following TBI.

Self-Location

As an authorship team, we represent the disciplines of music therapy (J.T. and J.B.) and neuropsychology (S.S.), with combined experience working across the lifespan from children to older adults, in acute, subacute, and community-based multidisciplinary rehabilitation programs as both clinicians and researchers. We are all white, cisgendered, heterosexual spouses and mothers. We (J.T. and J.B.) are extremely grateful for the opportunity to live and work on the unceded lands of Australia and have great respect for the traditional owners of these lands and their ongoing connection to country and culture.

Definitions and Background

Traumatic Brain Injury

Traumatic brain injury (TBI) is defined as an "alteration in brain function, or other evidence of brain pathology, caused by an external force" [3]. The distinguishing feature of a TBI is the damage to brain tissues that results from an external penetrating or non-penetrating force. Damage to brain tissue can also be a result of stroke, hypoxia, surgery, tumour, neurodevelopmental and/or neurodegenerative causes; however, this chapter will focus on traumatic injuries. The primary causes of TBI include motor vehicle accidents, falls, assault, and sporting injuries. The peak incidence of TBI is in the 15–25 year age range with males having the highest incidence, followed by young infants and elderly adults [4]. While TBI can occur across the lifespan, different mechanisms and recovery trajectories may be observed in children and older adults [5]. Prevalence of TBI is high, with an estimated 50–60 million people who experience a TBI each year worldwide [6]. Although most (90%) of these injuries are mild, the societal and economic burden of TBI is substantial.

TBI severity is generally defined by the Glasgow Coma Scale (GCS) score in the immediate period following the insult. Three categorisations of TBI severity are recognised: Mild TBI (GCS = 13–15), moderate TBI (GCS = 9–12), and severe TBI (GCS = 3–8) [4]. Severe TBI typically results in a loss of consciousness, and its duration, as well as length of post traumatic amnesia (PTA) [7], are negatively correlated with functional recovery and poorer prognosis [4]. Moderate and severe TBI commonly result in functional deficits and long-term sequelae, and we focus on these below.

Disorders of Consciousness (DoC)

Consciousness is defined as arousal *with* awareness [8, 9]. A severe TBI may interrupt the complex interplay of arousal and awareness, resulting in a disorder of consciousness (DoC). Arousal is the foundational tenet of consciousness and is assessed via the presence of a sleep-wake cycle that includes periods of eye opening [8, 10]. Awareness is a more complex phenomenon and includes an individual's perceptions, thoughts, and feelings. Arousal must be present for an individual to have awareness. Awareness is assessed via the presence of non-reflexive behavioural responses to external stimuli, including the ability to interact with others and/or the environment [10, 11].

DoC covers a spectrum of states including coma, unresponsive wakefulness syndrome (UWS) and minimally conscious state (MCS) [9]. DoC are seen more broadly in acquired brain injury, but in this chapter, we focus on TBI specifically. Coma is a state of sustained (>1 hour) unconsciousness in which both arousal and awareness are absent [9]. Individuals presenting in a coma do not display any meaningful response to external stimuli or interactions. During emergence from coma to UWS, arousal is present, but awareness remains absent. Individuals presenting in a MCS display arousal and emerging awareness, as evidenced by inconsistent but reproducible responses to external stimuli and interactions. MCS is assessed by an individual's ability to display a behavioural response to simple commands, visually fix and follow, communicate a yes/no response, and/or other contextually appropriate affective and behavioural responses. The presence of consistent functional communication and reliable object use are indicators of emergence to consciousness, noting that cognitive, motor, and/or language deficits may persist [9].

Post-traumatic Amnesia (PTA)

PTA is a phase of in recovery specific to severe TBI and represents a progression beyond DoC. It is characterised by an inability to orient to the environment or learn new information. The Westmead PTA Scale [12] is a standardised 12-item prospective measure of PTA that tests both normal day-to-day orientation and memory for new information as well as autobiographical memory.

TBI Sequelae and Rehabilitation Approaches

TBI causes a wide range of symptoms from physical deficits, including fatigue, pain, balance and motor problems, to communication impairments, and cognitive and emotional sequelae [6]. Clinical manifestations of TBI vary across patients depending on the type and severity of the injury and individual pre-injury factors [6]. Hemiplegia and muscle contractures are common following severe TBI, as are motor speech and language disorders. Due to the injury mechanism, which typically disrupts the functioning of large-scale brain networks via diffuse axonal injury and affects functioning of the frontal lobes, problems with executive functioning and attention are common [13]. High-level information processing that requires effective communication between distant brain areas is disrupted, and demanding tasks often result in cognitive fatigue [14]. Frontal executive dysfunction can lead to problems of initiation, planning and problem solving, inhibition and emotional regulation [13]. TBI triggers a complex neuroplastic process that can lead to either improvement or deterioration up to decades after the injury [15]. TBI also represents a risk factor for various neurological illnesses, such as epilepsy, stroke and neurodegenerative disease, including chronic traumatic encephalopathy [15].

Neural changes taking place after TBI are mediated by environmental factors that could contribute to either negative or positive neuroplasticity, with potential for a downward spiral if there is a lack of environmental enrichment [16]. Music is a powerful environmental factor: musical activities tap into several cognitive and motor functions and can be highly motivating and emotionally salient [17], which is important given the problems of initiation following TBI [18]. Additionally, music has the potential to support emotional adjustment and become a tool for self-expression.

Literature Review

The theoretical roots of music therapy practice in neurorehabilitation are grounded in neuroscience literature and supported in recent decades by brain imaging research into the neural mechanisms utilised for music perception and participation [19, 20]. This knowledge is used to design targeted music therapy interventions to stimulate neuroplastic changes in the injured brain, either by stimulating growth of new neural pathways, or by utilising unimpaired neural real estate to bypass and take over function of damaged areas. In this way, music therapy acts as a form of environmental enrichment, promoting widespread neural activation and using experience-dependent learning to drive activity-dependent neuroplasticity [21].

To maximize recovery outcomes, rehabilitation should ideally begin as soon as possible following severe brain injury [22]. Music therapy is ideally positioned for early rehabilitation because live music can be modified as needed by a trained therapist to regulate arousal and stimulate appropriate levels of engagement [23, 24]. There is also inherent temporal organization in music, which can stimulate neural reorganization for people in DoC needed to facilitate arousal, awareness, and orientation [25]. In addition to its organizing properties, music holds personal meaning and emotional salience for most people and is thus highly motivating, increasing patient engagement both musically and socially. In addition, many elements of cognition, such as attention, perception, memory, executive function, and learning are enhanced by positive affective states [26]. Thus, given the strong influence of music on emotion, music therapy can be used to target cognitive rehabilitation both directly and indirectly. Strong evidence also supports the use of music therapy interventions for motor rehabilitation and speech and language rehabilitation [27]. These areas and putative mechanisms for music-based interventions are covered in more depth in other chapters (see Chaps. 2, 3, 5, and 6).

The enriched sensory environment and multimodal stimulation provided by musical activities can also be utilised effectively for cognitive retraining and rehabilitation, even for patients with no prior musical experience or expertise. Research indicates that music supports executive functioning in healthy individuals across life span [28, 29] and highlights cognitive benefits of music therapy in TBI [30]. Temporal and structural cues inherent in music can be used therapeutically to assist attention control, learning, memory formation, and executive function. There is accumulating scientific evidence supporting particularly the efficacy of music therapy in the rehabilitation of executive functioning. Recently, in the first randomized controlled trial (RCT) exploring the use of neurological music therapy in moderate/severe TBI [31], music therapy was shown to support general executive functioning, set shifting and behavioral regulation [31, 32]. Furthermore, these gains were linked to specific volumetric and functional neuroplastic changes in the brain. Attention can be addressed by modifying the length, difficulty, or complexity of a musical activity. Songs can form an effective mnemonic device using rhythm and melody to organise, sequence, chunk, and recall verbal information. Rhythmic or melodic patterns can be learnt and replicated on an instrument beginning at an appropriate level of difficulty for each patient with progressive increases in complexity over time. Supported musical improvisation activities are useful, both to assess cognitive impairments and to provide opportunities for creativity and emotional expression [24, 31]. Therapeutic songwriting can also be used to enhance wellbeing and emotional adjustment, and support people to process changes in self-concept following acquired neurological injuries [33, 34].

Music therapy can play an important role in regulating arousal and awareness for people with DoC across the continuum from patients in coma through to low awareness states and PTA. The inability to orient to the environment or learn new information in PTA causes disorientation, confusion and susceptibility to overstimulation and frequently leads to agitated behaviours [35]. Use of familiar and personally salient music has been shown to increase orientation and decrease agitation for people in PTA [36] and in disorders of consciousness [37]. The capacity for music (that has been carefully selected and presented) to create a structured, safe, and more familiar environment, has been argued to explain its efficacy for enhancing orientation [38]. In support of the use of familiar and personally salient music, O'Kelly and colleagues [37] found that familiar music elicited responses indicative of arousal and selective attention more than disliked music. To maximise chances for the injured brain to rest and recover, usually only required care and essential therapy are recommended whilst a patient is in PTA [35]. Music therapy, however, provides a unique opportunity for direct intervention where musical stimuli can be regulated and used to actively influence behaviour (aiming to decrease agitation and increase orientation) without placing strenuous demands on the patient [39]. Receptive music therapy also provides the elements of positive distraction and focus, as well as opportunity for a social experience that does not require verbal interaction or complex cognitive processing. The familiarity and predictability of the music are key, as these assist both to alleviate anxiety and provide a reassuring, comforting experience with potential to stimulate pleasant memories. There is also the potential for the temporally regulated aspect of music to help organise a dysregulated and injured brain and to influence motor behaviour without conscious cognitive input [40]. All interventions should be individually targeted to maximise orientation while avoiding overstimulation and cognitive fatigue. Some individuals may benefit from receptive methods and others may respond positively to gentle but active engagement (e.g. singing along).

Music has unique affordance in the assessment and rehabilitation of individuals presenting with a DoC due to the frequently robust retention of hearing and auditory processing following severe TBI [41, 42]. This is particularly relevant in the context of high prevalence of visual impairment and language disorders for this population [42]. Further, current DoC guidelines highlight the importance of familiar, emotionally salient, self-referential stimuli in maximising early recovery of arousal and awareness [23]. Neuroscience evidence suggests that music stimulates subcortical networks involved in arousal and emotional processing, and also higher cortical frontal, parietal, and temporal regions that support awareness and higher cognitive processes [42]. Music is also able to simultaneously engage the neural networks involved in both internal self-referential awareness and external awareness, thus providing capacity to trigger an increase in consciousness of the self and the environment [41]. The use of music in the assessment of DoC is well documented [41]; for example, the MATADOC [43] is a useful tool to support the development of an individualised music therapy treatment program.

Evidence suggests that familiar music successfully increases consciousness and cognition in individuals presenting with a DoC, even when a behavioural response cannot be observed due to motor and cognitive deficits [37]. Given the points above, in our view live musical intervention should be prioritised when working with an individual presenting with a DoC [37, 44]. Live music can be adapted in the moment, contingent on any observed responses from the patient to encourage, extend or even suppress behavioural responses [41, 44]. This may include adaptations in the tempo, pitch, volume and/or complexity of the melody and harmonic accompaniment. For example, the music therapist may carefully introduce musical improvisation that incorporates the individual's name and has a tempo based on the patient's respiration rate. The contingent presentation of familiar songs also affords a predictable structure that can be adapted when performed live to adjust the level of sensory stimulation to meet individual patient needs. This reduces the potential for overstimulation and cognitive fatigue while maximising opportunities for engagement and meaningful stimulation to reduce the potential for under-stimulation.

Case Example 1: Using Music Therapy to Stimulate Responses in Early Consciousness Rehabilitation Following Severe TBI

The following case describes the contingent use of familiar song to support and stimulate increased responsiveness in a child presenting with an acute DoC following severe TBI. An analysis of the child's responses during music therapy sessions was undertaken from video footage. This case was part of a research study [44], and relevant ethics approval and parental consent were obtained.

Ten-year-old Evie sustained a severe TBI in a motor vehicle accident. She had a GCS of 6/15 at the scene of the accident and at the time of admission to the acute trauma hospital her pupils were fixed and dilated. Brain imaging indicated extensive areas of hematoma in the left parietal and temporal regions, multiple fragmented depressed skull fractures of the left occipital parietal, frontal, and temporal bones included areas of haemorrhage under the bones, midline shift, global diffuse axonal injury, and contrecoup injury to the right fronto-temporal regions.

The music therapy session detailed here occurred 25 days after Evie's TBI. She had emerged from coma but was not demonstrating eye fixing or following. She was producing some vocalisations through a speaking valve in her tracheostomy, but these did not appear to be intentional. She was not following verbal commands, had a right hemiplegia and presented with dystonic movements in her left arm. She had unsettled periods which corresponded with increased heart rate and blood pressure. Before her injury, Evie loved ABBA's music; therefore, ABBA songs were used in the music therapy session to provide a salient and familiar experience that was temporally structured to support an increase in neurological organisation and arousal. The songs were sung live by the music therapist so they could be adapted in response to any observed behaviours.

During the song 'Waterloo' Evie was observed to initially become still. The music therapist sang the song with a steady tempo, accompanying the lyrics with a simple strumming pattern on the guitar that emphasised the beat of the song. There was an observed decrease in Evie's dystonic movements, no vocalisations, and a relaxing of her facial muscles. During the final line of the verse, leading into the chorus, the music therapist slowed the tempo of the song, pausing the music before singing the chorus, to musically invite a response from Evie. During this musical pause, Evie was observed to draw a breath in, followed by increased muscle tension in her left arm, and pulling up of her legs. She then immediately exhaled and her whole body relaxed. The exhalation was also followed by a vocalisation upon commencement of the chorus, repeated during many renditions of the song. This was a significant response to the familiar song because Evie had not displayed purposeful responses to verbal commands during clinical observations and assessments. However, she repeatedly responded to familiar songs, at meaningful moments in the song. This indicates a level of intention and cognitive mediation from a child who was presenting with a DoC.

Case Example 2: Using Music Therapy to Reduce Agitation and Increase Orientation During Post-traumatic Amnesia

The following composite case example draws on principles for implementing music therapy during PTA outlined by Baker [36, 38] and Tamplin [39]. This includes live presentation of music that is familiar, predictable, and personally meaningful to the patient with modification of musical elements to reduce potential for overstimulation.

Sam, a 23-year-old non-binary person was hospitalised with a severe traumatic brain injury resulting from a motor vehicle accident. They had a GCS score of 7/15 on admission, with bilateral frontal contusions, subdural haemorrhage, and diffuse axonal injury. Sam was referred to music therapy for PTA management to increase orientation to the environment and reduce agitation. Patients are easily overstimulated due to hyper-arousal during PTA, so Sam's music therapy sessions were kept short, with no encouragement of conversation or active participation. As Sam was unable to remember new information or make sense of their environment, everything was unfamiliar, confusing, and frightening. This combination of confusion and fear was distressing for Sam and exacerbated their already heightened agitation levels. Therefore, one of the initial aims for music therapy was to reassure Sam and help them to feel safe. Reducing length of PTA improves prognosis, as the longer PTA symptoms last, the less likely is a full recovery to a premorbid level of function [7]. PTA recovery occurs most quickly when patients are relaxed and have lots of rest, therefore the purpose of Sam's music therapy was to encourage rest and sleep, as well as provide low stimulus live preferred music for emotional support and familiarity.

During music therapy assessment Sam presented as restless, confused, and agitated, pacing around the room, talking incoherently, and dressing/undressing repeatedly. Individual music therapy sessions were conducted twice weekly in Sam's room, using only voice and guitar to reduce potential for overstimulation and minimise safety risks for the therapist. A 2-min observation period (without intervention or interaction) was conducted before and after each session to monitor changes. Familiar, meaningful songs were identified in consultation with Sam's family. Songs were selected for therapy from Sam's preferred genres (heavy rock and beat-based dance music), and adapted using minimalist, ballad-style accompaniment at a slow, relaxing tempo with a quiet and consistent dynamic level. Only songs with positive lyrical content were used and priority was given to songs that were thought to have positive memory associations, such as their high school graduation song. The music therapist provided a simple explanation that she would be playing some songs and encouraged Sam to relax and listen. Sometimes when Sam was anxious and pacing, the therapist walked alongside Sam and sang using the Iso principle (where music that matches a patient's current mood or energy level is gradually altered to shift towards a desired state) to reduce arousal. Singing gently with a calming tone, she often brought Sam back to sit down on the bed, sometimes humming if lyrics were too stimulating. Verbal interaction was minimised, and Sam was not encouraged to actively participate in the music (however, sometimes they sang along spontaneously). The music was observed to have a regulating effect on Sam's agitated repetitive movements, such as pacing and tapping surfaces without purpose, where these transformed into more appropriate and intentional behaviours, such as swaying or tapping in time with the music. Further, the movements often ceased appropriately when the music finished. If the music seemed to exacerbate Sam's agitation or confusion, it was paused, and the session was terminated if necessary. Frequently, Sam fell asleep during sessions, and this was celebrated as a successful outcome of music therapy by the rehabilitation team. Sam's agitation was generally significantly lower in the post session observation period, and after 4 weeks Sam was assessed to have emerged from PTA.

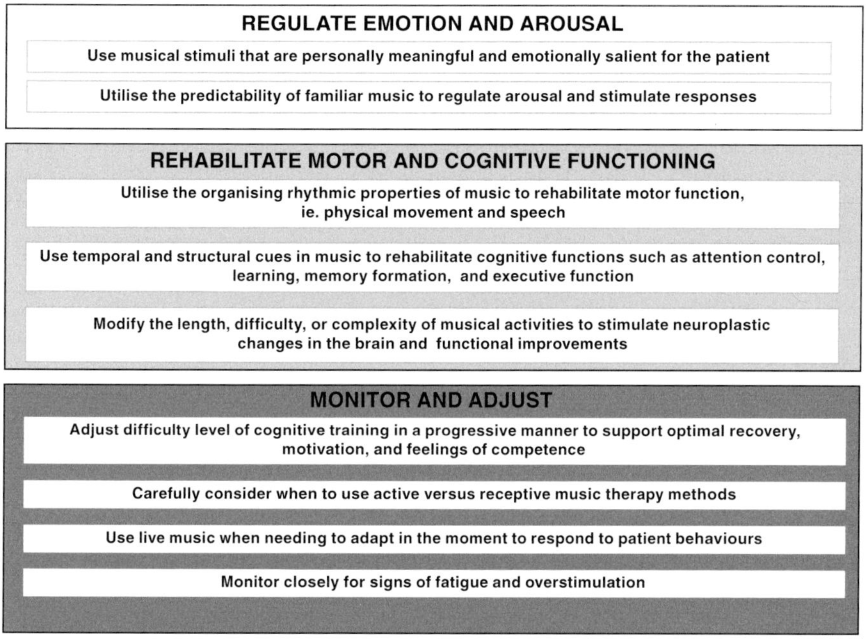

REGULATE EMOTION AND AROUSAL

Use musical stimuli that are personally meaningful and emotionally salient for the patient

Utilise the predictability of familiar music to regulate arousal and stimulate responses

REHABILITATE MOTOR AND COGNITIVE FUNCTIONING

Utilise the organising rhythmic properties of music to rehabilitate motor function, ie. physical movement and speech

Use temporal and structural cues in music to rehabilitate cognitive functions such as attention control, learning, memory formation, and executive function

Modify the length, difficulty, or complexity of musical activities to stimulate neuroplastic changes in the brain and functional improvements

MONITOR AND ADJUST

Adjust difficulty level of cognitive training in a progressive manner to support optimal recovery, motivation, and feelings of competence

Carefully consider when to use active versus receptive music therapy methods

Use live music when needing to adapt in the moment to respond to patient behaviours

Monitor closely for signs of fatigue and overstimulation

Fig. 4.1 Guidelines for delivery of music interventions in TBI/DoC

Guidelines for Delivery of Music Interventions in TBI/DOC

Ideally music therapy, like all rehabilitation therapies, should commence as early as possible following severe brain injury [22] and in consultation with the multidisciplinary team. This will maximise the potential for rehabilitation within the peak natural recovery time window when neuroplasticity is at its highest. However, music-based interventions commenced later in the rehabilitation journey are also beneficial given the long-term nature of recovery from severe brain injury [45]. Figure 4.1 below includes some guidelines to inform delivery of music-based interventions that have been presented in this chapter.

Conclusion and Future Considerations

This chapter presents an overview of the current research supporting music-based interventions in TBI rehabilitation. Case examples have been presented to highlight the value of individually targeted music interventions to maximise consciousness, orientation, and arousal regulation, and facilitate cognitive rehabilitation following severe TBI. Music is processed globally throughout the brain and current evidence indicates that some ability to meaningfully process and respond to music remains even following severe TBI [46]. Environmental factors are crucial in maintaining and extending function after TBI, which is a chronic condition with long trajectory of cognitive, behavioural, emotional, and motor complications. Music is a versatile, multimodal, and motivating stimulus, which provides unique therapeutic affordances. However, as with all stimulation, the use of music as therapeutic intervention should be carefully managed to avoid adverse effects including overstimulation and cognitive fatigue. This is particularly relevant in individuals who may have a limited repertoire of subtle behavioural responses due to their TBI. Individuals in the early stages of recovery from TBI may benefit from short but frequent music interventions rather than longer interventions that offer little opportunity for rest [24, 47].

A music therapist should actively engage with the multidisciplinary team in all aspects of TBI assessment and treatment to ensure the best possible patient care is provided. They may be able to offer unique insights into meaningful and repeated responses to salient stimuli that may not be observed during other therapies. Aligning music therapy goals with the treatment goals of the multidisciplinary team can support the incorporation of a rich range of techniques that will maximise the individual's engagement in music therapy; for example, a physiotherapist may support optimal positioning to maximise arousal for participation in music therapy. Finally, exploring the effects of music therapy and music-based interventions should be done through development of more rigorous studies that include biomarkers of response, such as electrocorticography signals, scalp EEG, MRI, functional near-infrared spectroscopy, and blood/CSF-based biomarkers (see Appendix A).

References

1. Ruttan L, Martin K, Liu A, Colella B, Green RE. Long-term cognitive outcome in moderate to severe traumatic brain injury: a meta-analysis examining timed and untimed tests at 1 and 4.5 or more years after injury. Arch Phys Med Rehabil. 2008;89(12):S69–76.
2. Sarajuuri J, Vink M, Tokola K. Relationship between late objective and subjective outcomes of holistic neurorehabilitation in patients with traumatic brain injury. Brain Inj. 2018;32(13–14):1749–57.
3. Menon DK, Schwab K, Wright DW, Maas AI. Position statement: definition of traumatic brain injury. Arch Phys Med Rehabil. 2010;91(11):1637–40.
4. Wagner AK, Franzese K, Weppner JL, Kwasnica C, Galang GN, Edinger J, et al. Traumatic brain injury. In: Braddom's physical medicine and rehabilitation. Elsevier; 2021. p. 916–953.e19.
5. Anderson V, Catroppa C, Morse S, Haritou F, Rosenfeld J. Functional plasticity or vulnerability after early brain injury? Pediatrics. 2005;116(6):1374–82.
6. Maas AI, Menon DK, Adelson PD, Andelic N, Bell MJ, Belli A, et al. Traumatic brain injury: integrated approaches to improve prevention, clinical care, and research. Lancet Neurol. 2017;16(12):987–1048.
7. Ponsford JL, Spitz G, McKenzie D. Using post-traumatic amnesia to predict outcome after traumatic brain injury. J Neurotrauma. 2016;33(11):997–1004.
8. Di Perri C, Thibaut A, Heine L, Soddu A, Demertzi A, Laureys S. Measuring consciousness in coma and related states. World J Radiol. 2014;6(8):589–97.
9. Giacino JT, Katz DI, Schiff ND, Whyte J, Ashman EJ, Ashwal S, et al. Practice guideline update recommendations summary: disorders of consciousness: report of the guideline development, dissemination, and implementation Subcommittee of the American Academy of Neurology; the American Congress of Rehabilitation Medicine; and the National Institute

on Disability, Independent Living, and Rehabilitation Research. Arch Phys Med Rehabil. 2018;99(9):1699–709.
10. Laureys S, Schiff ND. Coma and consciousness: paradigms (re) framed by neuroimaging. NeuroImage. 2012;61(2):478–91.
11. Giacino JT, Fins JJ, Laureys S, Schiff ND. Disorders of consciousness after acquired brain injury: the state of the science. Nat Rev Neurol. 2014;10(2):99–114.
12. Shores EA, Marosszeky JE, Sandanam J, Bachelor J. Preliminary validation of a clinical scale for measuring the duration of post-traumatic amnesia. Med J Aust. 1986;144(11):569–72.
13. Ham TE, Sharp DJ. How can investigation of network function inform rehabilitation after traumatic brain injury? Curr Opin Neurol. 2012;25(6):662–9.
14. Ramage AE, Tate DF, New AB, Lewis JD, Robin DA. Effort and fatigue-related functional connectivity in mild traumatic brain injury. Front Neurol. 2019;9:1165.
15. Wilson L, Stewart W, Dams-O'Connor K, Diaz-Arrastia R, Horton L, Menon DK, et al. The chronic and evolving neurological consequences of traumatic brain injury. Lancet Neurol. 2017;16(10):813–25.
16. Tomaszczyk JC, Green NL, Frasca D, Colella B, Turner GR, Christensen BK, et al. Negative neuroplasticity in chronic traumatic brain injury and implications for neurorehabilitation. Neuropsychol Rev. 2014;24(4):409–27.
17. Blood AJ, Zatorre RJ. Intensely pleasurable responses to music correlate with activity in brain regions implicated in reward and emotion. Proc Natl Acad Sci U S A. 2001;98(20):11818–23.
18. Marin RS, Wilkosz PA. Disorders of diminished motivation. J Head Trauma Rehabil. 2005;20(4):377–88.
19. Chatterjee D, Hegde S, Thaut M. Neural plasticity: the substratum of music-based interventions in neurorehabilitation. NeuroRehabilitation. 2021;48(2):155–66.
20. Martínez-Molina N, Siponkoski S-T, Kuusela L, Laitinen S, Holma M, Ahlfors M, et al. Resting-state network plasticity induced by music therapy after traumatic brain injury. Neural Plast. 2021;2021:1–18.
21. Sihvonen AJ, Särkämö T. Clinical and neural predictors of treatment response to music listening intervention after stroke. Brain Sci. 2021;11(12):1576.
22. Seel RT, Douglas J, Dennison AC, Heaner S, Farris K, Rogers C. Specialized early treatment for persons with disorders of consciousness: program components and outcomes. Arch Phys Med Rehabil. 2013;94(10):1908–23.
23. Schnakers C, Magee WL, Harris B. Sensory stimulation and music therapy programs for treating disorders of consciousness. Front Psychol. 2016;7:297.
24. Baker FA, Tamplin J. Music therapy methods in neurorehabilitation: a clinician's manual. London, Philadelphia: Jessica Kingsley Publishers; 2006.
25. Thaut MH. Rhythm, music, and the brain: scientific foundations and clinical applications. Taylor & Francis; 2008.
26. Tyng CM, Amin HU, Saad MN, Malik AS. The influences of emotion on learning and memory. Front Psychol. 2017;8:1454.
27. Magee WL, Clark IN, Tamplin J, Bradt J. Music interventions for acquired brain injury (review). Cochrane Database Syst Rev. 2017;(1):CD006787.
28. Bugos JA, Perlstein WM, McCrae CS, Brophy TS, Bedenbaugh PH. Individualized piano instruction enhances executive functioning and working memory in older adults. Aging Ment Health. 2007;11:464–71.
29. Moreno S, Bialystok E, Barac R, Schellenberg EG, Cepeda NJ, Chau T. Short-term music training enhances verbal intelligence and executive function. Psychol Sci. 2011;22(11):1425–33.
30. Thaut MH, Gardiner JC, Holmberg D, Horwitz J, Kent L, Andrews G, et al. Neurologic music therapy improves executive function and emotional adjustment in traumatic brain injury rehabilitation. Ann N Y Acad Sci. 2009;1169:406–16.
31. Siponkoski S-T, Martínez-Molina N, Kuusela L, Laitinen S, Holma M, Ahlfors M, et al. Music therapy enhances executive functions and prefrontal structural neuroplasticity after traumatic brain injury: evidence from a randomized controlled trial. J Neurotrauma. 2020;37(4):618–34.

32. Siponkoski S-T, Koskinen S, Laitinen S, Holma M, Ahlfors M, Jordan-Kilkki P, et al. Effects of neurological music therapy on behavioural and emotional recovery after traumatic brain injury: a randomized controlled cross-over trial. Neuropsychol Rehabil. 2021;32:1356–88.
33. Tamplin J, Baker FA, MacDonald RAR, Roddy C, Rickard NS. A theoretical framework and therapeutic songwriting protocol to promote integration of self-concept in people with acquired neurological injuries. Nord J Music Ther. 2015;25:111–33. https://doi.org/10.1080/08098131.2015.1011208.
34. Baker FA, Tamplin J, Rickard N, Ponsford J, New PW, Lee YEC. A therapeutic songwriting intervention to promote reconstruction of self-concept and enhance wellbeing following brain or spinal cord injury: pilot randomised controlled trial. Clin Rehabil. 2019;33(6):1045–55.
35. Parker TD, Rees R, Rajagopal S, Griffin C, Goodliffe L, Dilley M, et al. Post-traumatic amnesia. Pract Neurol. 2022;22(2):129–37.
36. Baker FA. The effects of live, taped and no music on people experiencing posttraumatic amnesia. J Music Ther. 2001;38(3):170–92.
37. O'Kelly J, James L, Palaniappan R, Fachner J, Taborin J, Magee WL. Neurophysiological and behavioural responses to music therapy in vegetative and minimally conscious states. Front Hum Neurosci. 2013;7(884):1–15.
38. Baker FA. Rationale for the effects of music therapy on agitation and orientation levels of people in posttraumatic amnesia. Nord J Music Ther. 2001;10(1):32–41.
39. Tamplin J. Adults in critical care. In: Allen J, editor. Guidelines for music therapy practice in adult medical care. Phoenixville, PA: Barcelona Publishers; 2013. p. 62–83.
40. Thaut MH. Rhythm, music and the brain: scientific foundations and clinical applications. New York: Taylor & Francis Group; 2007.
41. Magee WL. Music in the diagnosis, treatment and prognosis of people with prolonged disorders of consciousness. Neuropsychol Rehabil. 2018;28(8):1331–9.
42. Perrin F, Castro M, Tillmann B, Luauté J. Promoting the use of personally-relevant stimuli for investigating patients with disorders of consciousness. Front Psychol. 2015;6:1102.
43. Magee WL, Siegert RJ, Taylor SM, Daveson BA, Lenton-Smith G. Music therapy assessment tool for awareness in disorders of consciousness (MATADOC): reliability and validity of a measure to assess awareness in patients with disorders of consciousness. J Music Ther. 2016;53(1):1–26.
44. Bower J, Catroppa C, Grocke D, Shoemark H. Music therapy for early cognitive rehabilitation post-childhood TBI: an intrinsic mixed methods case study. Dev Neurorehabil. 2014;17(5):339–46.
45. Teasell R, Mehta S, Pereira S, McIntyre A, Janzen S, Allen L, et al. Time to rethink long-term rehabilitation management of stroke patients. Top Stroke Rehabil. 2012;19(6):457–62.
46. Bower J, Magee WL, Catroppa C, Baker FA. The neurophysiological processing of music in children: a systematic review with narrative synthesis and considerations for clinical practice in music therapy. Front Psychol. 2021;12(1053):615209.
47. Sánchez AM, Pool J, Bower J, Paasch V, Magee W. Best practice recommendations for using music with children and young people with disorders of consciousness. Music Med. 2023;15(1) https://doi.org/10.47513/mmd.v15i1.885.

Music for Movement Disorders

Yuko Koshimori, Kyurim Kang, Kerry Devlin, and Alexander Pantelyat

Introduction

This chapter will provide an overview of music-based interventions (MBI) for various movement disorders, with an emphasis on music therapy approaches. Much of the research in this area has been done in Parkinson disease (PD), which is one of the most common movement disorders, though we also cover available (though limited) literature demonstrating use of MBI for/with patients with atypical parkinsonism and Huntington disease. We aim to highlight current understandings of key neurophysiological mechanisms pertinent to MBI in movement disorders, such as improving movement automaticity through external rhythmic cueing, as well as address the scope of music therapy approaches and MBI utilized with this patient community. We will also highlight knowledge gaps, linking these to future research directions.

Self-Locations

Dr. Yuko Koshimori is a staff scientist at Music and Health Science Research Collaboratory of University of Toronto and specializes in multimodal neuroimaging investigations of neurological conditions including investigating imaging correlates

Y. Koshimori
Music and Health Science Research Collaboratory (MaHRC), Faculty of Music, University of Toronto, Toronto, ON, Canada
e-mail: yuko.koshimori@utoronto.ca

K. Kang · K. Devlin · A. Pantelyat (✉)
Johns Hopkins Center for Music and Medicine, Department of Neurology, Johns Hopkins University School of Medicine, Baltimore, MD, USA
e-mail: kkang19@jhmi.edu; Kdevlin5@jh.edu; apantel1@jhmi.edu

© The Author(s), under exclusive license to Springer Nature Switzerland AG 2023
K. Devlin et al. (eds.), *Music Therapy and Music-Based Interventions in Neurology*, Current Clinical Neurology,
https://doi.org/10.1007/978-3-031-47092-9_5

of clinical symptoms of Parkinson's disease (PD), exploring novel imaging markers of PD and traumatic brain injury, and uncovering the neural mechanisms underlying the beneficial effects of rhythm-based therapeutic interventions such as Rhythmic Auditory Stimulation (RAS®) for PD.

Dr. Kyurim Kang is a postdoctoral research fellow and a neurologic music therapist at the Johns Hopkins Center for Music and Medicine. Her research focuses on translational and clinical research by understanding neural mechanisms underlying the effects of music interventions using brain imaging techniques and integrating this understanding with neurologic music therapy approaches for application in clinical settings. As both a researcher and a clinician, she values the interconnection between research and clinical practices, "research-based clinical practices as well as patients-oriented/centered research" to maximize the benefits of music to individuals who need the support.

Kerry Devlin is a white, disabled music therapist who has collaborated with therapy participants for nearly a decade in wide ranging settings and systems, including private practice, community-based settings, school systems, and now, a major hospital system where she is both a provider and a patient. Kerry teaches courses centered on disability, music therapy clinical practice, inclusive music education practices, and community music therapy. Her doctoral work explores the impact of therapist perspective in and on music therapy clinical practice and pedagogy.

Dr. Pantelyat is a movement disorders neurologist specializing in neurodegenerative diseases that affect movement and cognitive function. His clinical research centers on music-based interventions for neurodegenerative disorders and biomarker development and validation for parkinsonian disorders. Dr. Pantelyat has played violin since age 7 and it has been a major part of his life.

Definitions/Background

Movement Disorders

Movement disorders are a group of neurological illnesses that impact motor function and are typically subdivided on the basis of phenomenological (visual) observation of individuals' positions and movements. These include hypokinetic (involving reduced movement amplitude and/or speed) and hyperkinetic (involving excessive movement) disorders [1]. Hypokinetic movement disorders include parkinsonian disorders (Parkinson disease being the most common), while hyperkinetic movement disorders include chorea (Huntington disease being the most common), tremor, ballism, tics, stereotypies ataxias and myoclonias [2]. An individual can exhibit multiple types of movement disorders simultaneously and the nature of the movement disorder within an individual can evolve over time, depending on the underlying disease process.

Rhythm for Movement Disorders

Rhythmic Auditory Stimulation (RAS®)

RAS® is a Neurologic Music Therapy (NMT®) technique that uses an auditory rhythmic cue to entrain gait to a specific rhythm [3]. It is applied clinically by identifying the patient's baseline cadence first, and then providing rhythmic cues with a metronome and/or live music at progressively faster or slower frequencies (usually 5–10% increments) to improve the patient's gait. This anticipatory time cue is effective as both an immediate entrainment stimulus, providing rhythmic cues during movement, as well as a facilitating stimulus to achieve and maintain a more functional gait pattern [3, 4]. The following sections provide an overview of the potential neural mechanisms and principles of RAS® (see also Chap. 2).

Neural Mechanisms of Auditory Rhythm and Music

To date, investigations of brain changes induced by rhythmic auditory cueing and rhythm-based interventions have been found only in PD (see Appendix B). First, neurophysiological studies suggest that the rhythmic auditory cueing and RAS®-assisted intervention can induce enhancement of alpha and beta event-related desynchronization (ERD) and event-related synchronization (ERS) in sensory-motor areas and in the subthalamic nucleus (STN), as well as increased alpha and beta coupling in sensory-motor and auditory-motor areas, which are accompanied with improvement of motor functions [5–7]. These findings may be of importance in PD as abnormal oscillatory activities, particularly in the beta frequency band, are well-established electrophysiological phenomena of PD [8] with clinical relevance [9, 10]. Second, some studies showed that rhythmic auditory cueing recruits the cerebellar activity [11, 12]. Following a rhythm-based rehabilitation program, PD participants showed increased cerebellar glucose activity and improved repetitive motor behaviors including gait and finger tapping [10]. Cerebellar abnormality is also part of PD pathophysiology associated with tremor [13] as well as postural-instability-and-gait-disorder motor subtype [14]. Lastly, one study suggested that dopamine might be involved in auditory-motor synchronization, demonstrating an association between dopaminergic denervation and poor synchronization performance [15]. However, as this study did not investigate dopaminergic responses during auditory-motor synchronization, further studies are needed to determine the role of dopamine. In young healthy individuals, the dopaminergic response in the ventral striatum was reduced during a auditory-motor synchronization task compared to that during a continuation task, suggesting that the application of rhythmic auditory cueing may spare dopaminergic resources [16]. In another study with young healthy individuals, music-induced intense pleasure enhanced dopaminergic responses in the caudate nucleus and in the nucleus accumbens. These were associated with the anticipation of an abstract reward, and the pleasurable experience itself, respectively (Table 5.1) [17].

Table 5.1 Rhythmic auditory cueing studies in Parkinson disease with assessments of brain function

Study design/techniques	Participants/characteristics	Task or intervention	Primary outcome measures	Primary results (with rhythmic auditory cueing)
Experimental				
LFPs [74]	• 16 PD on med with DBS electrode bilaterally implanted in STN (age 61 ± 4 years; DD, 12 ± 4 years, 1 left-handed, 1 woman)	• Experimental tasks: synchronize steps, watching each heel strike displayed on the video (42 s containing 21 left and 21 right heel strikes) with and without auditory cueing (metronome sound at interval of 1 s) • In active control task: watch the video without moving and to think of anything unrelated to walking such as holidays or upcoming plans	• Step-to-cue difference, step interval durations, step timing variability, and step interval variability • Beta power modulation	• Significant reduction in step timing variability with auditory cueing, higher beta modulation, and higher beta rebound and faster beta decrease associated with longer interval of steps
LFPs [75]	• 7 PD on med with DBS electrode bilaterally implanted in STN (mean age 63 ± 6.5 years; 3 women; DD, 6–18 years)	• Move one hand or foot between two dots separated by 30 cm, cued by metronome cues of 1.6, 3.2, and 4.8 Hz during normal DBS, 80% DBS amplitude, no DBS	• Action tremor	• With auditory cueing and DBS combined, significant reduction in the number of extremities showing action tremor
MEG [76]	• 21 PD on med (age 66.3 ± 7.4 years, 12 women) on med • 23 HCs (age 65.8 ± 7.7 years, 12 women)	• Right index finger tapping with rhythmic (every 1 s) and non-rhythmic (every 1 s ± 200 ms) acoustic burst stimuli (a total of 6 sequences of 30 s separated by a 5-s rest period)	• Mean distance to cue • Oscillatory neuronal activity within 15–80 Hz	• No significant group difference in finger tapping performance • During the rhythmic condition, greater activation in bilateral frontal, supplementary and primary motor areas and occipital regions in HC compared to PD and greater activity in bilateral angular and supramarginal gyri and left rolandic operculum • During rhythmic condition compared to non-rhythmic condition, greater activity in temporal, parietal, and motor areas in HC, and greater activity in left angular and right middle frontal gyri in PD

fMRI [77]	• 27 PD on med who could walk independently and showed improvement of gait with RAS (age, 74.6 ± 6.8 years; 16 women; DD, 6.3 ± 4.5 years; H&Y II:III, 11:16; MMSE-Japanese, 27.6 ± 2.3) • 25 HC (age 71.8 ± 5.5 years, 12 women; MMSE-J 27.9 ± 2.2)	• Experimental tasks: white noise with gait imagery and rhythmic stimuli (electrically synthesized beep sound presented at 100 beats/min) with gait imagery • Control tasks: white noise at rest and rhythmic stimuli at rest	• Brain activation	• RAS vs white noise: reduced activity in the left parietal operculum in PD and no difference between the two conditions in HC • Gait imaginary vs rest: no difference in brain activity regardless of auditory stimuli in PD and increased activity in the left superior temporal gyrus with RAS compared to with white noise in HC
fMRI [11]	• 14 PD on med (age range, 51–81; 5 women) • 13 HC (age range, 51–78; 8 women)	• 1 Hz finger tapping with/without RAS • 4 Hz finger tapping with/without RAS	• Signed error (difference in timing between tap and onset of the stimulus tone) • Behavioral freezing (presence of large pauses during tapping) • Intra- and inter-network connectivity	• Significant reduction in variability of finger tapping across the group with RAS • With 4 Hz RAS, significant reduction in freezing in PD • No group difference in signed error • During 1 Hz tapping, greater intra-network connectivity in auditory, salience, motor/IFG, basal ganglia/thalamus, and motor/cerebellar networks in HC compared to PD • Greater inter-network connectivity in three pairs of networks: between auditory and executive control networks; between executive control and motor/cerebellar network; and between auditory and visual networks in PD compared to HC with and without RAS

(continued)

Table 5.1 (continued)

Study design/techniques	Participants/characteristics	Task or intervention	Primary outcome measures	Primary results (with rhythmic auditory cueing)
[11C]-DTBZ PET [78]	• 28 PD (age, 67.3 ± 7.5 years; 5 women; DD, 5.2 ± 3.2 years; H&Y I–III; UPDRS motor: 21.6 ± 7.2 (off), 19.7 ± 7.2 (on); MoCA: 25.9 ± 3.1 (off), 26.3 ± 2.7 (on); MMSE: 28.9 ± 1.1 (off), 28.8 ± 1.2 (on)	• Synchronized right or left index finger tapping to tone sequences of either 500, 1000 or 1500 ms time intervals while ON levodopor placebo pill	• Accuracy and variability of synchronization • Striatal dopaminergic denervation	• Less accurate synchronizing to 500 ms time interval, compared to the 1000 and 1500 ms time intervals • No effects of medication state, the hand used on synchronization accuracy or variability • No association between dopaminergic denervation and synchronization accuracy or variability • Three subgroups (based on degree of denervation) had partially paralleled behavioral differences in synchronization accuracy
Intervention RCT with EEG [79]	50 PD on med: • RAS (n = 25; age, 70 ± 8 years; 9 women; DD, 10 ± 3 years; H&Y, 3 ± 1; MMSE, 26 ± 3) • No RAS (n = 25; age, 73 ± 8 years; 9 women; DD, 9.3 ± 3 years; H&Y, 3 ± 1; MMSE, 25 ± 3)	• Individual treadmill gait training with or without RAS supervised by physiotherapists (30 min, once/day, 5 days/week for 8 weeks) • Walk along with beat (superimposed salient high-pitch bell sound) of music "angel elsewhere" presented with the lyrics, which reaches a target music tempo of ~120 bpm	• FGA; UPDRS; BBS; FES; 10MWT; TUG; GQI • ERD/ERS magnitude and functional connectivity in alpha and beta bands	• Significant improvement of FES, FGA, and UPDRS and increase in GQI in RAS • Significant improvement of BBS and TUG in both groups • Greater EEG power increase in ERD and ERS in alpha and beta bands within the frontal and centro-parietal areas in RAS • Greater fronto-centroparietal/temporal connectivity in beta bands in RAS • Greater increase in the fronto-centroparietal and fronto-temporal beta connectivity correlated with greater FGA improvement in both groups combined

RCT with EEG [80][a]	50 PD on med: • RAS (n = 25; age, 70 ± 8 years; 9 women; DD, 10 ± 2 years; H&Y, 3 ± 1; MMSE, 26 ± 3) • No RAS (n = 25; age 73 ± 8 years; 9 women; DD, 9.3 ± 3 years; H&Y, 3 ± 1; MMSE, 25 ± 3)	• Individual treadmill gait training with or without RAS supervised by physiotherapists (30 min, once/day, 5 days/week for 8 weeks) • Walk along with beat (superimposed salient high-pitch bell sound) of music "Animals Everywhere" presented with the lyrics, which reaches a target music tempo of ~120 bpm	• EEG amplitude/activation	• Greater reduction in frontal area activation and more potentiation in the centroparietal areas activation in RAS compared to no RAS • Gait-cycle specificity and reduction of cerebellar activity observed only in RAS • Improvement of gait (combined FGA and GQI) associated with decreased activity within frontal and cerebellar regions, and with increased activity within central and parietal regions
Case-control study with EEG [81]	20 PD on med: • DBS (n = 10; age, 62 ± 5 years; 4 women; DD, 15 ± 2 years; H&Y, 3 (2.5–3); MMSE, 26 (25–27); UPDRS-III ON, 17 (14–21); UPDRS III OFF, 32 (27–44)) • No DBS (n = 10: age, 62 ± 4 years; 5 women; DD, 14 ± 2 years; H&Y, 2.5 (2.5–2.9); MMSE, 27(27); UPDRS-III ON, 28 (27, 28); UPDRS III OFF, 37 (34–45))	• Treadmill gait training with or without RAS on med (30 min, once/day, 4 days/week for a month) • Simple two-accent metronome sounds of 120 bpm or the maximum tolerable bpm	• UPDRS-III ON and OFF; BBS; FES; 10MWT; TUG; ACE-R • ERSP in relation to RAS provision in the alpha and beta frequency ranges	• BBS, FES, and ACE-R significantly improved in both groups • Greater improvement in UPDRAS OFF, TUG, and 10MWT in DBS group • Greater remodulation of sensorimotor alpha and beta oscillations associated with the gait cycle in both groups • Significant association between beta power percent change (decrease) within motor programing ROI and 10MWT percent change (increase) in both groups combined

(continued)

Table 5.1 (continued)

Study design/techniques	Participants/characteristics	Task or intervention	Primary outcome measures	Primary results (with rhythmic auditory cueing)
Case study with MEG [7]	3 PD on med: • Participant 1: 62 yo, woman; H&Y, 2.5; UPDRS III, 23 • Participant 2: 70 yo, man, H&Y, 3; UPDRS III, 52 • Participant 3: 72 yo, H&Y, 2; UPDRS III, 24	• 15 sessions of somatosensory-related NMT techniques (3 times/week for 5 consecutive weeks) provided by a NMT-certified music therapist • Each session consisting of bimanual exercises using a keyboard, castanets and miscellaneous objects to strengthen fine motor muscles • Each finger movement cued by either a metronome or beats produced by the therapist playing musical instrument	• UPDRS-III; GPT; finger-thumb opposition task • Evoked power and functional connectivity of frequency ranging 15–80 Hz in left auditory and primary motor cortices during cued finger tapping	• Improvement in one or more areas of fine motor functions for either dominant hand or both hands in all participants • Possibly coinciding increase in beta-range evoked power • Increased functional connectivity between primary auditory and motor cortices

CCT with [¹⁸F]-FDG PET [12]	• 9 PD on med (age, 61 ± 5; 4 women; DD, 3–8 years; H&Y, 1–2.5 stages; UPDRS, 12–45) • 5 HCs (age, 63 ± 4; 3 women)	• Gait and repetitive arm training consisting of 20 1-h session over 4 weeks • Gait and/or upper limb movements cued by a metronome at rates of 30–150 bpm • Upper limb movements cued by a metronome at rates of 0.5 and 4 Hz	• Gait parameters (velocity; step length; cadence; variability) • Finger tapping (frequency; variability) • Glucose metabolism	• Higher variability in gait and finger tapping and less velocity in PD compared to HC before training • In PD, variability in gait and finger tapping significantly decreased to the similar level of HC after training • Hypometabolism in right parietal and temporal lobes, left temporal lobe, left frontal lobe, and left cerebellum (culmen of anterior lobe) in PD compared to HC before training. • Increase in glucose metabolism in right cerebellum including anterior lobe and dentate nucleus, as well as right parietal (BA39) and temporal lobes after training compared to before training in PD

Mean ± SD. Median (iqr)

ACE-R Addenbrooke's cognitive examination–revised, *BBS* Berg balance scale, *CCT* controlled clinical trial, *DBS* deep brain stimulation, *DD* duration of disease, *EEG* electroencephalogram, *ERD* event-related desynchronization, *ERS* event-related synchronization, *ERSP* event–related spectral perturbation, *FES* falls efficacy scale, *FGA* functional gait assessment, *fMRI* functional magnetic resonance imaging, *GPT* grooved Pegboard test, *GQI* gait quality index, *HC* healthy control, *H&Y* Hoehn and Yahr, *LFP* local field potential, *MEG* magnetoencephalography, *MMSE* mini mental state examination, *MoCA* Montreal cognitive assessment, *NMT* neurologic music therapy, *PET* positron emission tomography, *PD* Parkinson's disease, *RAS* rhythmic auditory stimulation, *RCT* randomized control trial, *STN* subthalamic nucleus, *TUG* timed up-and-go test, *yo* years old, *10MWT* 10-m walking test

[a] Secondary EEG analysis of the study by Naro et al. (2020)

Literature Review

Neurologic Music Therapy (NMT®)

The use of rhythmic beats, familiar pattern music, playing musical instruments, and other musical components can facilitate movement exercises for patients with movement disorders. RAS® has been widely and effectively utilized in rehabilitation settings for patients with movement disorders (especially for PD) to improve their gait parameters, balance, and reduce freezing of gait (FoG) based on rhythmic entrainment and anticipation of timing principles (see review [18–21]). Duration of gait improvements using RAS® has ranged from immediate to 6 months [22–24]. In addition to investigating the effects of RAS® in isolation, there is an increasing body of research focused on comprehensively exploring the combined effects of RAS® with other training modalities in PD (such as treadmill exercises and physiotherapy) [5, 25, 26] and elucidating the underlying neural mechanisms involved [5]. For example, Calabrò et al. (2019) discovered that an 8-week training program involving treadmill gait training with RAS resulted in improved gait performance in PD. Furthermore, the study revealed a more pronounced increase in EEG power during specific phases of the gait cycle when RAS® was utilized, along with a greater enhancement in frontocentroparietal/temporal electrode connectivity [5].

Rhythmic cueing has also been used to improve upper limb movements in PD [27]. In this study, the participants' hand dexterity was measured using the Purdue Pegboard Test[1] while listening to three different tempi of rhythmic cues (baseline tempo, 10% faster of the baseline, and 20% faster of the baseline). They found the highest scores with the fastest tempo they provided (20% incremented over the baseline) compared to the baseline tempo. This study demonstrated promising results of using rhythmic cues for movements that are less cyclic and rhythmic than gait.

Using rhythmic cueing, familiar musical patterns (i.e., patterned sensory enhancement, PSE®) and musical instrument playing (i.e., Therapeutic Instrumental Musical Performance, TIMP®), a study conducted by [7] found that fine motor function significantly improved after 15 TIMP® sessions (using keyboards, castanets and other objects to strengthen fine motor muscles) accompanied by rhythmic cues provided by a metronome or beats produced by a musical instrument. In addition, Magnetoencephalography (MEG)[2] measurements revealed significant changes in cortical beta band activity and increases in functional connectivity between the auditory and motor cortex with cueing [7]. This neuroimaging study during the intervention suggests potential mechanisms of musical interventions for fine motor impairments. Furthermore, previous research showed that 30 h of piano and music

---

[1] Purdue pegboard test (PPT) is used to measure unimanual and bimanual finger and hand dexterity (Tiffin et al. [82]).

[2] Magnetoencephalography (MEG) measures the magnetic fields produced by electrical activity in the brain.

theory training over 10 days enhanced performance on Stroop[3] and musical self-efficacy measurements [28].

There is limited evidence for the use of NMT® and music-based interventions in movement disorders other than PD (including Huntington's disease, HD and atypical parkinsonism, APD). Thaut et al. (1999) investigated if HD patients can modulate gait velocity in response to an auditory rhythmic cue (e.g., normal, slower, and faster self-paced walking). The results showed that patients were able to significantly modulate their gait velocity during rhythmic metronome cueing, but not during music [29]. In contrast, Delval et al. (2008) found that auditory cues did not improve gait parameters in HD patients, and this may be due to a greater degree of attention difficulties than in PD patients [30].

Atypical parkinsonian (APD) disorders include progressive supranuclear palsy (PSP), corticobasal syndrome (CBS), multiple system atrophy (MSA) and dementia with Lewy bodies (DLB). To date, only a handful of pilot studies tested the effectiveness of RAS® in APD. Wittwer et al. (2019) explored a home-based music-cued therapy program in five participants with PSP. After the 4-week program (eight sessions), all five participants reported being satisfied with the therapy and clinically improved in their gait velocity and stride length variability [31]. Another RAS® study in APD, Pantelyat et al. (2022) found significantly increased cadence (steps/minute) during and after RAS® compared to baseline in the overall cohort, though patients with DLB experienced a worsening gait pattern, possibly owing to a greater degree of cognitive dysfunction that interfered with the ability to walk and process external cues simultaneously [32]. Recently, a novel clinical intervention for CBS has been proposed to improve their hand and arm coordination using PSE® and TIMP® combined with transcranial direct current stimulation (tDCS) [33].

Given the challenges of care access in patients with movement disorders (cost, travel time, inconvenience), home-based interventions have been an active area of investigation. A therapeutic device for gait has been recently developed using RAS® components/mechanisms (e.g., MedRhythms, see Chap. 14) and is being investigated in stroke [34], PD and multiple sclerosis rehabilitation. Mainka and his team developed an application ("CuraSwing") was designed to improve arm swing in PD based on RAS® and sensor-based musical feedback [35]. They analyzed gait parameters including arm swings in three different conditions: (1) normal walking; (2) focused swinging of the more affected arm; and (3) with musification of arm swing feedback (using "CuraSwing'), and they found an immediate effect of musification on arm swing and other gait parameters in PD. The use of these two therapeutic devices has sparked the idea that both RAS® and closed-loop musical feedback can be effective tools for promoting the gait patterns in patients with PD. It is important to note that safety is a paramount concern for patients utilizing home-based interventions, including therapeutic devices. Several neurologic music therapists have reported the safety concerns and difficulty with some assessments, especially for the gait and other motor movements [36] and safety assessments should be included as outcome measures in future studies.

---

[3] Stroop test is used to assess the ability to inhibit cognitive interference.

Rhythm, the key element of sensorimotor domain rehabilitation, has been known to induce expectation about sentence prosody, and enhance sentence processing [37]. Specifically, rhythmic speech cueing has been used to facilitate speech intelligibility in people with PD [38] and speech production/accuracy [39]. More detailed information about speech disorders is in Chap. 6.

As an author team, we feel it important to note that while the literature exploring NMT® interventions tends to focus on functional performance (e.g., improving gait), this way of working also holds possibilities for emotional and social wellbeing. Emerging literature has begun exploring the impact of NMT® interventions on psychosocial domains, examples of which include virtual group NMT® for individuals with PD to improve apathy and depression [40] and once-weekly NMT® sessions in a subacute neurorehabilitation context to support engagement in physical exercises *and* to improve mood and motivation [41]. Furthermore, Kang et al. (2023) are presently conducting a study that examines anxiety and emotional states (including arousal, pleasure, and dominance) as well as motor outcomes in patients with corticobasal syndrome during the implementation of PSE® and TIMP® for improving upper extremity performance [33]. This investigation takes into account the significance of psychological factors, aiming to evaluate their impact on the effectiveness of these therapeutic techniques. The importance of building trust between patient and therapist, cyclic assessment of the patient's changing needs, and creating opportunities for emotional expression when utilizing NMT® techniques cannot be overstated.

Integrative Music Therapy Approaches

While NMT® appears to be the most heavily studied music therapy approach within neurorehabilitation contexts, this reflects one way of working among many. Raglio (2015) offers their conceptualization of music therapy clinical practice with patients diagnosed with PD as being either "rehabilitative" and "relational" [42]. They frame rehabilitative approaches to therapy as grounded in application of concrete musical tasks to improve targeted aspects of functioning (e.g., NMT®), which can be differentiated from relational approaches that emphasize improvisational music making and relationship-building (e.g., music-centered approaches). Bruscia [43] also describes three major ways of thinking in and about music therapy practice at large—outcome-oriented (e.g., NMT®), experience-oriented (e.g., Nordoff-Robbins Music Therapy), and context or ecologically-oriented (e.g., community music therapy)—emphasizing that each approach uses music differently to respond to different therapeutic needs [44, 45].

The reality of music therapy in any clinical context is that the work—and the people with whom we work—rarely exist in binaries. Instead, clinicians make decisions in response to a patient's primary and emergent needs, which may involve combining approaches or consciously choosing one way of working over another in service of the patient's needs and desires [46]. Given this, we advocate for an approach to music therapy clinical practice that invites multiple ways of working to co-exist in complementary fashion acknowledging that each offers its own unique set of characteristics and interventions that can be accessed fluidly in service of the patient and their needs. Without this flexibility, Jochims (2004) cautions that neurorehabilitation solely focused on one aspect of human experience (e.g., motor functioning alone) may fail to meet the changing needs of the patient, arguing instead for a holistic approach to therapy that addresses physical, sensory, cognitive, communication, social, and emotional needs in parallel [47].

Music therapists have described their work with patients with lived experience of PD and HD in both individual and group music therapy contexts, with special emphasis given to the centrality of the therapeutic relationship to the therapy process [43, 48, 49].The literature in this area tends to emphasize individual cases and study designs with small sample sizes that reflect wide-ranging approaches to clinical practice, with some systematic reviews (e.g., [42, 50, 51]) indicating the potential benefits of music therapy on mood, quality of life, motor functioning, and communication in this patient community. Music therapists have documented use of clinical approaches including improvisation, songwriting, receptive music experiences, music and movement, singing, and digital music making to support wide-ranging motor (e.g., gait, mobility, balance) and non-motor (e.g., quality of life, mood, cognition, social connections) goals of relevance to patients with movement disorders [42, 49–57]. Integration of music therapy and MBI into end-of-life care (see Chap. 9) for individuals with movement disorders like Lewy Body Dementia and PD has also been explored as an avenue to facilitate reminiscence, family bonding, affirmation of personhood, and pain management [58, 59].

When collaborating with individuals with PD, some music therapists have described the importance of flexible, integrative approaches that shift to meet a patient's needs and desires as they are revealed throughout the therapy process. Panebianco and Lotter's (2019) work with 'Susan', a 70-year-old patient with PD, demonstrates the ways in which different music therapy approaches can be used in tandem across eight 60-min music therapy sessions [49]. During her therapy process, Susan engaged in active music therapy (e.g., structured instrument play and vocal work; vocal and instrumental improvisation), music and movement (e.g., relaxation, breathing, and stretching; dancing; walking), and receptive music therapy (e.g., music listening; music and imagery; visual art creation) experiences. Thematic analysis of session videos, the therapist's reflections, and Susan's own reflections revealed the following themes connected to the potential benefits of music therapy in the context of Susan's life: 1) motivation to engage in therapy, 2) control and regulation when playing instruments and walking, 3) intentionality and focus when making music, 4) confidence in her music, 5) spontaneity and creativity, 6) emotional support within the therapeutic relationship, and 7) generalization of skills learned in music therapy to other environments (p. 89).

Music therapists working with patients diagnosed with HD have described similar approaches to clinical practice, with music therapy sessions focusing on psychosocial well-being (see Chap. 16), enhancing emotional expression, communication, reducing social isolation, improving quality of life, and family-centered care [52, 60, 61]. Patients themselves have articulated the potential role of music therapy in their lives, suggesting that psychosocial support, building confidence/self-esteem, cultivating self-expression, providing opportunities for autonomy, and a developing sense of security could be integral to music therapy sessions targeting quality of life in HD [61]. Individual cases have emphasized the use of music as a medium for communication, such as engaging patients in songwriting processes to share lived experiences with care partners [52] and using receptive music listening experiences to address behavioral and mood-related symptoms [60]. Though van Bruggen-Rufi et al. (2018) were unable to confirm anecdotal benefits of music therapy reported in case study literature (e.g., impact on communication outcomes in advanced HD), they suggest that different outcome measures emphasizing quality of life could be more meaningful for future research given the complexity of working with patients experiencing late-stage disease [61].

Music-Based Interventions: Music Medicine

Presented below are several examples of Music Medicine MBI used for movement disorders outside of Music Therapy. These examples involve work done at the Johns Hopkins Center for Music and Medicine, with which 3 of the 4 chapter authors are affiliated; it should be noted that outstanding work is being done in this area at other centers and is omitted here simply due to chapter length requirements.

Group singing has been investigated in PD in multiple studies that have considered motor and non-motor outcomes. Voice loudness, swallow function, quality of life, mood and anxiety are among the outcomes that have been positively impacted by group singing in PD, with interventions typically occurring 1–2 times/week and no evidence for superiority of a particular frequency [50, 62]. Choirs for individuals with PD around the world have been regularly meeting for over a decade (see for example the Parkinson Voice Project: https://parkinsonvoiceproject.org/). For more information on this topic, the reader is referred to Devlin et al. (2019) [50], Butala et al. [63], Good et al. (2023) [62] and Lee and Dvorak [64].

Drumming has been investigated for PD [65] and HD in several small studies [66, 67]. A controlled pilot study in PD (Drum-PD) found significant quality of life improvements in a group of 8 patients who underwent instructor-led West African drumming sessions as a group twice per week for 6 weeks. These improvements became non-significant 6 weeks after the end of intervention, indicating that continued drumming sessions are needed to maintain benefit. In HD, an open-label study of individual at-home drumming (15 min/day, 5 times/week for 2 months) in 5 patients found improvements in executive function and the white matter microstructure of the genu corpus callosum (connecting prefrontal cortices of the cerebral hemispheres) as assessed by MRI [66]. The same group later performed a controlled study (8 HD and 9 controls) using the same approach and found improvements in drumming quality and executive function; however, these improvements did not correlate with microstructural changes in white matter [67].

Another example of a Music Medicine approach involved 26 PD participants who were randomized 1:1 to 6 weeks of twice per week guitar classes taught by conservatory faculty or to usual care control (GuitarPD study); the usual care group began guitar classes 6 weeks after the study began and outcomes were assessed every 6 weeks for 18 weeks [68]. Over 90% of participants completed the study and group analysis showed clinically and statistically significant improvements in self-reported mood immediately after completion of guitar classes. There was also a trend ($p = 0.07$) toward overall quality of life improvement in the same time frame. No significant improvements in hand dexterity were observed and improvements in mood and quality of life did not persist across all participants at weeks 12 and 18. As in the Drum-PD study above and in line with other behavioral interventions, this suggests that continued practice is necessary to sustain longer-term improvements.

Case Examples

Case 1: PSE®/TIMP® for Corticobasal Syndrome (CBS) (Contributed by Kyurim Kang)

Sarah (renamed due to confidentiality), a 60-year-old woman, had been diagnosed with CBS 3 years prior to the beginning of the therapy session in June 2022. CBS is a condition with limited treatment options, prompting Sarah to express eagerness and readiness to explore any available options for management. CBS patients typically exhibit asymmetrical involuntary movements, including rigidity, tremor, dystonia, and often apraxia. In addition, nonmotor deficits such as cognitive and language impairments are commonly observed.

Upon assessment, Sarah demonstrated rigidity, dystonia, and apraxia predominantly on her left side (her non-dominant side), while her cognitive abilities were preserved. Considering her specific needs, the intervention aimed to target hand dexterity, arm coordination, and shoulder range of motion.

The sessions were scheduled for 30 min, twice a week, for a total of three consecutive weeks (a total of 6 sessions) The duration and frequency of the sessions were determined based on clinical considerations and the available time frame.

During each session, the primary interventions utilized were PSE® and TIMP®. The session began with a warm-up routine involving musical patterns, which included activities such as making circles with different body parts (e.g., head rotation, shoulder rounding, wrist movements) and synchronized breathing exercises. These activities aimed to engage Sarah in rhythmic patterns and prepare her for the subsequent interventions. The interventions focused on the affected side initially, gradually progressing to include both sides to observe and address any coordination differences. Activities such as bicep curls, open/close hand exercises, and supination/pronation movements were performed in sync with the musical patterns. Additionally, various musical instruments such as the mini keyboard, cabasa, hand drums, and maracas were incorporated to facilitate engagement and sensorimotor stimulation.

Reported outcomes from Sarah were predominantly anecdotal but provided valuable insights. She expressed her perception that the rhythm of the interventions felt smoother and more comfortable. However, interestingly, the objective measurement scores did not reflect the same improvement. She mentioned that rhythmic cueing helped her facilitate and better understand her movements in response to music, effectively targeting her challenges.

Case 2: Virtual Support Group (Contributed by Kerry Devlin)

"What do you need today?" I ask the Zoom room as our check-in conversation begins to quiet. I scan all 8 boxes on my computer screen, each containing a different face or group of faces—15 participants total. We sit in silence for a moment before one person offers, "What if we do something where music can just…wash over us? And we can just feel it together?" Another person quickly unmutes with an energetic, "Yes! That reminds me of…" and begins to sing Judy Garland without warning. *"Never could carry a tune, never knew where to start. You came along when everything was wrong and put a song in my heart."* Another person unmutes and joins in, *"Dear, when you smiled at me, I heard a melody…"* The Zoom room is alive with music and connection as the singing fades to applause. Then, our soloist says, "Hold on—why don't we hear from someone who hasn't had the chance to share yet today. What do *you* need?"

This moment captures what is, for me, the essence of our monthly support group—that it has evolved into a space rooted in community care and collective desire. There are no "entry criteria" to join other than having lived experience with a neurologic diagnosis as a patient or care partner, resulting in a rich peer network comprised of folx at every stage of diagnosis with many different diagnoses. No two people come with the same lived experience, yet every participant's lived experience seems to intersect in some way when it comes to shared challenges, joys, and experiences navigating healthcare. Every session is structured differently because it responds to whoever attends and their needs in the moment, and the group has decided its structure all the way down to when we meet and how often (once per month). In many ways, the group is led by its own participants in dialogic fashion—who sometimes ask for suggestions from the facilitating music therapists and sometimes suggest their own ways of being together by initiating music through singing, sharing recordings, and song requests.

Another interesting layer to this group dynamic is the fact that it is co-facilitated by two music therapists working from two different paradigms: NMT® (Kyu) and striving toward a disability justice informed anti-oppressive practice (Kerry). Some experiences we have implemented in response to group needs include NMT® techniques (e.g., TIMP®; PSE®; Therapeutic Singing (TS®) when structured movement or vocal work that can be practiced at home are requested, more flexible music and movement experiences (e.g., dancing freely to songs to rekindle joyful connection with the body), receptive music listening (e.g., music and relaxation experiences; song sharing with live and recorded music), song discussion (e.g., lyric analysis), and verbal processing in response to music experiences and group conversation. Though I can only reflect through lens of my own experience, I find that this integrative approach has created new opportunities for participants to make choices about their care in ways not afforded by other healthcare encounters, centering their autonomy and desires as central to the therapy process.

Important note: This vignette is written from my (Kerry's) felt experience as a co-facilitator. Identifying details and descriptions of group experiences above have been altered and reimagined to protect confidentiality.

Conclusion and Future Directions

Taken together, prior studies suggest that rhythmic auditory cueing can influence brain activity in cortical and subcortical regions as well as in the cerebellum, and through this modulate different neurochemical systems. These multi-level modulations induced by rhythmic auditory cueing can be possible because auditory areas have close functional connectivity with motor-associated areas in the central nervous system [69–71]. More studies are needed to further uncover the neural mechanisms underlying the beneficial effects of rhythmic auditory cueing on movement disorders including PD. Although some studies demonstrated correlations between brain changes and improved motor functions, due to a lack of control data, it is still unknown whether these brain changes are compensatory for or modulation of PD pathophysiology. Control data are crucial to (1) confirm the abnormalities found in PD, (2) compare brain changes post-intervention and (3) brain activity during auditory-motor synchronization between controls and PD. Furthermore, more neurophysiological data in a large sample size are needed to investigate entrainment and cross-frequency coupling in distributed areas of the brain and a relationship between these neuromodulatory effects and clinical symptoms. More studies are also needed in PD with deep brain stimulation (DBS) and other movement disorders such as essential tremor and dystonia, as this can provide unique opportunities to investigate the relationship in neurophysiological activities between cortical and DBS implant sites as well as the relationship between neurophysiological activity and the clinical symptoms. In addition, it would be interesting to further investigate whether auditory-motor synchronization and auditory-motor synchronization cued by auditory rhythm embedded in pleasure-inducing music modulate residual dopaminergic function in people with dopamine deficiency.

Studies published to date indicate that individuals with movement disorders (in particular PD, which is a common disease that has been studied far more than other movement disorders) have much to gain from MBI. NMT® utilizing RAS® has shown much promise for gait rehabilitation and other relevant outcomes in movement disorders. Integrative music therapy practices, such as experience and context-dependent approaches, merit further exploration in those with PD, HD and other movement disorders. This is a particularly relevant future area of study given a recent systematic review and meta-analysis of existing literature confirmed efficacy of MBI for improvement of some motor symptoms in PD but found limited evidence for the impact of MBI on *non-motor* symptoms (e.g., cognitive and emotional disturbances) [72]. Because integrative approaches may involve different therapeutic goals from those of NMT®, different methods (e.g., qualitative and mixed methods analysis) and different outcomes (e.g., ecological momentary assessments of mood, anxiety and stress levels) need to be assessed for these types of studies [73]. We see promise in studying and implementing combined music therapy approaches for patients with movement disorders such as PD because the disease process involves many non-motor aspects that may adversely impact quality of life.

Although the evidence for MBI in movement disorders (particularly PD) is growing, many fundamental questions remain to be addressed. Among them, we highlight: (1) whether specific Music Therapy approaches, Music Medicine approaches, or their combination are most relevant for particular movement disorders; (2) optimal dosage and frequency of musical/rhythmic stimulation; (3) the importance of individualization of MBI for patient-driven outcome optimization; (4) outcome comparison for expanded access to home-based vs clinic-based vs. community-based MBI; (5) comparison of in-person vs telehealth-based MBI approaches; (6) investigation of the impact of MBI on motor *and* non-motor symptoms; (7) investigation of MBI coupled with non-invasive brain stimulation such as tDCS; and (8) evaluation of neurophysiological responses before, *during* and after MBI to better understand underlying mechanisms. Additionally, our literature review of research on MBI for movement disorders outside of PD indicated that there is broad opportunity to develop MBI for hyperkinetic movement disorders such as chorea, tics and dystonia and to explore MBI for atypical parkinsonian disorders. We conclude by highlighting the tension that at times exists between the type of MBI evidence needed by public research funding agencies and health insurance payors (e.g., randomized controlled trials focused on specific treatment comparisons and outcomes generalized across participant groups) and the type of evidence that emerges from an integrative, individualized approach to MBI delivery often favored by music therapists and those they work with. We believe that future work needs to utilize both types of approaches and their combination.

References

1. Jellinger KA. Neuropathology and pathogenesis of extrapyramidal movement disorders: a critical update—I. Hypokinetic-rigid movement disorders. J Neural Transm. 2019;126(8):933–95.
2. Sanger TD, Chen D, Fehlings DL, Hallett M, Lang AE, Mink JW, et al. Definition and classification of hyperkinetic movements in childhood. Mov Disord. 2010;25(11):1538–49.
3. Thaut M, Hoemberg V. Handbook of neurologic music therapy. Oxford: Oxford University Press; 2014.
4. Thaut M. Rhythm, music, and the brain: scientific foundations and clinical applications. Routledge; 2013.
5. Calabrò RS, Naro A, Filoni S, Pullia M, Billeri L, Tomasello P, et al. Walking to your right music: a randomized controlled trial on the novel use of treadmill plus music in Parkinson's disease. J Neuroeng Rehabil. 2019;16(1):1–14.
6. Fischer P, Chen CC, Chang YJ, Yeh CH, Pogosyan A, Herz DM, et al. Alternating modulation of subthalamic nucleus beta oscillations during stepping. J Neurosci. 2018;38(22):5111–21.

7. Buard I, Dewispelaere WB, Thaut M, Kluger BM. Preliminary neurophysiological evidence of altered cortical activity and connectivity with neurologic music therapy in Parkinson's disease. Front Neurosci. 2019;13:105.
8. Boon LI, Geraedts VJ, Hillebrand A, Tannemaat MR, Contarino MF, Stam CJ, et al. A systematic review of MEG-based studies in Parkinson's disease: the motor system and beyond. Hum Brain Mapp. 2019;40(9):2827–48.
9. Chen R, Berardelli A, Bhattacharya A, Bologna M, Chen KHS, Fasano A, et al. Clinical neurophysiology of Parkinson's disease and parkinsonism. Clin Neurophysiol Pract. 2022;7:201–27.
10. Neumann WJ, Kühn AA. Subthalamic beta power-unified Parkinson's disease rating scale III correlations require akinetic symptoms. Mov Disord. 2017;32(1):175–6.
11. Braunlich K, Seger CA, Jentink KG, Buard I, Kluger BM, Thaut MH. Rhythmic auditory cues shape neural network recruitment in Parkinson's disease during repetitive motor behavior. Eur J Neurosci. 2019;49(6):849–58.
12. del Olmo MF, Arias P, Furio MC, Pozo MA, Cudeiro J. Evaluation of the effect of training using auditory stimulation on rhythmic movement in Parkinsonian patients–a combined motor and [18F]-FDG PET study. Parkinsonism Relat Disord. 2006;12(3):155–64.
13. Zhong Y, Liu H, Liu G, Zhao L, Dai C, Liang Y, et al. A review on pathology, mechanism, and therapy for cerebellum and tremor in Parkinson's disease. npj Parkinsons Dis. 2022;8(1):82.
14. Basaia S, Agosta F, Francia A, Cividini C, Balestrino R, Stojkovic T, et al. Cerebro-cerebellar motor networks in clinical subtypes of Parkinson's disease. npj Parkinsons Dis. 2022;8(1):113.
15. Miller NS, Kwak Y, Bohnen NI, Müller ML, Dayalu P, Seidler RD. The pattern of striatal dopaminergic denervation explains sensorimotor synchronization accuracy in Parkinson's disease. Behav Brain Res. 2013;257:100–10.
16. Koshimori Y, Strafella AP, Valli M, Sharma V, Cho S, Houle S, et al. Motor synchronization to rhythmic auditory stimulation (RAS) attenuates dopaminergic responses in ventral striatum in young healthy adults: [11C]-(+)-PHNO PET study. Front Neurosci. 2019;13:106.
17. Salimpoor VN, Benovoy M, Larcher K, Dagher A, Zatorre RJ. Anatomically distinct dopamine release during anticipation and experience of peak emotion to music. Nat Neurosci. 2011;14(2):257–62.
18. Ghai S, Ghai I, Schmitz G, Effenberg AO. Effect of rhythmic auditory cueing on parkinsonian gait: a systematic review and meta-analysis. Sci Rep. 2018;8(1):1–19.
19. Lim I, van Wegen E, de Goede C, Deutekom M, Nieuwboer A, Willems A, et al. Effects of external rhythmical cueing on gait in patients with Parkinson's disease: a systematic review. Clin Rehabil. 2005;19(7):695–713.
20. Spaulding SJ, Barber B, Colby M, Cormack B, Mick T, Jenkins ME. Cueing and gait improvement among people with Parkinson's disease: a meta-analysis. Arch Phys Med Rehabil. 2013;94(3):562–70.
21. Zhang M, Li F, Wang D, Ba X, Liu Z. Mapping research trends from 20 years of publications in rhythmic auditory stimulation. Int J Environ Res Public Health. 2023;20(1):215.
22. Capato TT, de Vries NM, IntHout J, Ramjith J, Barbosa ER, Nonnekes J, et al. Multimodal balance training supported by rhythmic auditory stimuli in Parkinson disease: effects in freezers and nonfreezers. Phys Ther. 2020;100(11):2023–34.
23. Hausdorff JM, Lowenthal J, Herman T, Gruendlinger L, Peretz C, Giladi N. Rhythmic auditory stimulation modulates gait variability in Parkinson's disease. Eur J Neurosci. 2007;26(8):2369–75.
24. Kadivar Z, Corcos DM, Foto J, Hondzinski JM. Effect of step training and rhythmic auditory stimulation on functional performance in Parkinson patients. Neurorehabil Neural Repair. 2011;25(7):626–35.
25. De Luca R, Latella D, Maggio MG, Leonardi S, Sorbera C, Di Lorenzo G, et al. Do patients with PD benefit from music assisted therapy plus treadmill-based gait training? An exploratory study focused on behavioral outcomes. Int J Neurosci. 2020;130(9):933–40.
26. Naro A, Pignolo L, Sorbera C, Latella D, Billeri L, Manuli A, et al. A case-controlled pilot study on rhythmic auditory stimulation-assisted gait training and conventional physiotherapy

26. in patients with Parkinson's disease submitted to deep brain stimulation. Front Neurol. 2020;11:794.
27. Fan W, Li J, Wei W, Xiao SH, Liao ZJ, Wang SM, et al. Effects of rhythmic auditory stimulation on upper-limb movements in patients with Parkinson's disease. Parkinsonism Relat Disord. 2022;101:27–30.
28. Bugos JA, Lesiuk T, Nathani S. Piano training enhances Stroop performance and musical self-efficacy in older adults with Parkinson's disease. Psychol Music. 2021;49(3):615–30.
29. Thaut MH, Miltner R, Lange HW, Hurt CP, Hoemberg V. Velocity modulation and rhythmic synchronization of gait in Huntington's disease. Mov Disord. 1999;14(5):808–19.
30. Delval A, Krystkowiak P, Delliaux M, Blatt JL, Derambure P, Destée A, et al. Effect of external cueing on gait in Huntington's disease. Mov Disord. 2008;23(10):1446–52.
31. Wittwer JE, Winbolt M, Morris ME. A home-based, music-cued movement program is feasible and may improve gait in progressive supranuclear palsy. Front Neurol 2019;10 [cited 2022 May 16]. Available from: https://www.frontiersin.org/article/10.3389/fneur.2019.00116.
32. Pantelyat A, Dayanim G, Kang K, Turk B, Pagkatipunan R, Huenergard SK, et al. Rhythmic auditory cueing in atypical parkinsonism: a pilot study. Front Neurol. 2022;13:1018206.
33. Kang K, Stenum J, Roemmich RT, Heller NH, Jouny C, Pantelyat A. Neurologic music therapy combined with EEG-tDCS for upper motor extremity performance in patients with corticobasal syndrome: study protocol for a novel approach. Contemp Clin Trials. 2023;125:107058.
34. Hutchinson K, Sloutsky R, Collimore A, Adams B, Harris B, Ellis TD, et al. A music-based digital therapeutic: proof-of-concept automation of a progressive and individualized rhythm-based walking training program after stroke. Neurorehabil Neural Repair. 2020;34(11):986–96.
35. Mainka S, Scholl A, Warmerdam E, Gandor F, Maetzler W, Ebersbach G. The power of musification: sensor-based music feedback improves arm swing in Parkinson's disease. Mov Disord Clin Pract. 2021;8(8):1240–7.
36. Cole LP, Henechowicz TL, Kang K, Pranjić M, Richard NM, Tian GL, et al. Neurologic music therapy via telehealth: a survey of clinician experiences, trends, and recommendations during the COVID-19 pandemic. Front Neurosci. 2021;15:648489.
37. Cason N, Astésano C, Schön D. Bridging music and speech rhythm: rhythmic priming and audio–motor training affect speech perception. Acta Psychol. 2015;155:43–50.
38. Thaut MH, McIntosh KW, McIntosh GC, Hoemberg V. Auditory rhythmicity enhances movement and speech motor control in patients with Parkinson's disease. Funct Neurol. 2001;16(2):163–72.
39. Wambaugh JL, Martinez AL. Effects of rate and rhythm control treatment on consonant production accuracy in apraxia of speech. Aphasiology. 2000;14(8):851–71.
40. Shah-Zamora D, Anderson S, Barton B, Fleisher JE. Virtual group music therapy for apathy in Parkinson's disease: a pilot study. J Geriatr Psychiatry Neurol. 2023:089198872311767. https://doi.org/10.1177/08919887231176755.
41. Thompson N, Bloska J, Abington A, Masterson A, Whitten D, Street A. The feasibility and acceptability of neurologic music therapy in subacute neurorehabilitation and effects on patient mood. Brain Sci. 2022;12(4):497.
42. Raglio A. Music therapy interventions in Parkinson's disease: the state-of-the-art. Front Neurol. 2015;6:185.
43. Dietz L. The integral role of the therapeutic relationship within a neuroscience-informed approach to music therapy practice: a philosophical inquiry. Master's thesis. Montreal, Quebec: Concordia University; 2022. Available from: https://spectrum.library.concordia.ca/id/eprint/990469/1/Dietz_MACATS%28MT%29_W2022.pdf.
44. Bruscia. Defining music therapy. 3rd ed. Barcelona Publishers; 2016.
45. American Music Therapy Association. AMTA-Pro podcast series. 2012 [cited 2023 Jun 13]. Ken Bruscia: AMTA's 2011 Sears Distinguished Lecture Series Speaker. Available from: https://amtapro.musictherapy.org/?p=797.
46. Devlin K, Meadows A. Considering case formulation and decision-making processes in music therapy assessment and treatment planning. Nord J Music Ther. 2023;32:1–17.

47. Jochims S. Music therapy in the area of conflict between functional and psychotherapeutic approach within the field of neurology/neurorehabilitation. Nord J Music Ther. 2004;13(2):161–71.
48. Devlin K. Making meaning out of life-altering illness: a clinician's reflections on music therapy sessions with a couple processing the impact of corticobasal syndrome. In: Conference presentation presented at International Association for Music and Medicine; 2020. Available from: https://iammonline.com/iamm2020/.
49. Panebianco C, Lotter C. Shifting 'stuckness' in Parkinson's disease: a music therapy case study. J Music Arts Afr. 2019;16(1–2):77–97.
50. Devlin K, Alshaikh JT, Pantelyat A. Music therapy and music-based interventions for movement disorders. Curr Neurol Neurosci Rep. 2019;19(11):1–13.
51. Machado Sotomayor MJ, Arufe-Giráldez V, Ruíz-Rico G, Navarro-Patón R. Music therapy and Parkinson's disease: a systematic review from 2015–2020. IJERPH. 2021;18(21):11618.
52. Brandt M, Helmers A. F65 'it became difficult to communicate with him but music opened him up for us' family music therapy for a patient with Huntington's disease, a case description. J Neurol Neurosurg Psychiatry. 2022;93(Suppl 1):A60–1.
53. DeLucia D. The experiences of an individual with Parkinson's engaging in a relationship-based, improvisational music therapy group. Master's thesis. New York: Malloy University; 2022. Available from: https://digitalcommons.molloy.edu/cgi/viewcontent.cgi?article=1124&context=etd.
54. Kogutek D, Ready E, Holmes JD, Grahn JA. Evaluating note frequency and velocity during improvised active music therapy in clients with Parkinson's disease. In: LaGasse B, Baker F, editors. J Music Ther. 2023;60(1):36–63.
55. Pacchetti C, Mancini F, Aglieri R, Fundarò C, Martignoni E, Nappi G. Active music therapy in Parkinson's disease: an integrative method for motor and emotional rehabilitation. Psychosom Med. 2000;62(3):386–93.
56. Spina E, Barone P, Mosca LL, Forges Davanzati R, Lombardi A, Longo K, et al. Music therapy for motor and nonmotor symptoms of Parkinson's disease: a prospective, randomized, controlled, single-blinded study. J Am Geriatr Soc. 2016;64(9):e36–9.
57. Stegemoller E, Forsyth E, Patel B, Elkouzi A. Group therapeutic singing improves clinical motor scores in persons with Parkinson's disease. BMJ Neurol Open. 2022;4(2):e000286.
58. Chaplain R. Exploring the hinterland: the development of a person-centered music therapy method for a hospice patient with Lewy body dementia. Master's thesis. Lesley University; 2023. Available from: https://digitalcommons.lesley.edu/expressive_theses/703.
59. Dimaio L. Music therapy entrainment: a humanistic music therapist's perspective of using music therapy entrainment with hospice clients experiencing pain. Music Ther Perspect. 2010;28(2):106–15.
60. Hu M. A survey of music therapists who work with clients with Huntington's disease. Master's thesis. Ohio University; 2022. Available from: http://rave.ohiolink.edu/etdc/view?acc_num=ohiou1649267420459963.
61. Van Bruggen-Rufi M, Vink A, Achterberg W, Roos R. Improving quality of life in patients with Huntington's disease through music therapy: a qualitative explorative study using focus group discussions. Nord J Music Ther. 2018;27(1):44–66.
62. Good A, Earle E, Vezer E, Gilmore S, Livingstone S, Russo FA. Community choir improves vocal production measures in individuals living with Parkinson's disease. J Voice. 2023; https://doi.org/10.1016/j.jvoice.2022.12.001.
63. Butala A, Li K, Swaminathan A, Dunlop S, Salnikova Y, Ficek B, Portnoff B, Harper M, Vernon B, Turk B, Mari Z, Pantelyat A. Parkinsonics: a randomized, blinded, cross-over trial of group singing for motor and nonmotor symptoms in idiopathic parkinson disease. Parkinsons Dis. 2022;2022:4233203.
64. Lee SJ, Dvorak AL. Therapeutic Group Singing for Individuals with Parkinson's Disease: A Conceptual Framework. Music Therapy Perspectives. 2023;41(2):178–186.
65. Pantelyat A, Syres C, Reichwein S, Willis A. DRUM-PD: the use of a drum circle to improve the symptoms and signs of Parkinson's disease (PD). Mov Disord Clin Pract. 2016;3(3):243–9.

66. Metzler-Baddeley C, Cantera J, Coulthard E, Rosser A, Jones DK, Baddeley RJ. Improved executive function and callosal white matter microstructure after rhythm exercise in Huntington's disease. J Huntingtons Dis. 2014;3(3):273–83.
67. Casella C, Bourbon-Teles J, Bells S, Coulthard E, Parker GD, Rosser A, et al. Drumming motor sequence training induces apparent myelin remodelling in Huntington's disease: a longitudinal diffusion MRI and quantitative magnetization transfer study. J Huntingtons Dis. 2020;9(3):303–20.
68. Bastepe-Gray S, Wainwright L, Lanham DC, Gomez G, Kim JS, Forshee Z, et al. GuitarPD: a randomized pilot study on the impact of nontraditional guitar instruction on functional movement and well-being in Parkinson's disease. Parkinsons Dis. 2022;2022:1061045.
69. Grahn JA, Brett M. Rhythm and beat perception in motor areas of the brain. J Cogn Neurosci. 2007;19(5):893–906.
70. Chen JL, Penhune VB, Zatorre RJ. Listening to musical rhythms recruits motor regions of the brain. Cereb Cortex. 2008;18(12):2844–54.
71. Thaut MH. The discovery of human auditory–motor entrainment and its role in the development of neurologic music therapy. Prog Brain Res. 2015;217:253–66.
72. Lee H, Ko B. Effects of music-based interventions on motor and non-motor symptoms in patients with Parkinson's disease: a systematic review and meta-analysis. IJERPH. 2023;20(2):1046.
73. Edwards E, St Hillaire-Clarke C, Frankowski DW, Finkelstein R, Cheever T, Chen WG, et al. NIH music-based intervention toolkit: music-based interventions for brain disorders of aging. Neurology. 2023;100:868–78.
74. Fischer P, Chen CC, Chang YJ, et al. Alternating modulation of subthalamic nucleus Beta oscillations during stepping. J Neurosci. 2018;38(22):5111–21.
75. Heida T, Wentink EC, Zhao Y, Marani E. Effects of STN DBS and auditory cueing on the performance of sequential movements and the occurrence of action tremor in Parkinson's disease. J Neuroeng Rehabil. 2014;11:135.
76. Buard I, Dewispelaere WB, Teale P, et al. Auditory entrainment of motor responses in older adults with and without Parkinson's disease: an MEG study. Neurosci Lett. 2019;708:134331.
77. Nishida D, Mizuno K, Yamada E, Hanakawa T, Liu M, Tsuji T. The neural correlates of gait improvement by rhythmic sound stimulation in adults with Parkinson's disease—a functional magnetic resonance imaging study. Parkinsonism Relat Disord. 2021;84:91–7.
78. Miller NS, Kwak Y, Bohnen NI, Muller ML, Dayalu P, Seidler RD. The pattern of striatal dopaminergic denervation explains sensorimotor synchronization accuracy in Parkinson's disease. Behav Brain Res. 2013;257:100–10.
79. Calabro RS, Naro A, Filoni S, et al. Walking to your right music: a randomized controlled trial on the novel use of treadmill plus music in Parkinson's disease. J Neuroeng Rehabil. 2019;16(1):68.
80. Naro A, Pignolo L, Bruschetta D, Calabro RS. What about the role of the cerebellum in music-associated functional recovery? A secondary EEG analysis of a randomized clinical trial in patients with Parkinson disease. Parkinsonism Relat Disord. 2022;96:57–64.
81. Calabro RS, Naro A, Cimino V, et al. Improving motor performance in Parkinson's disease: a preliminary study on the promising use of the computer assisted virtual reality environment (CAREN). Neurol Sci. 2020;41(4):933–41.
82. Tiffin J, Asher EJ. The Purdue Pegboard: norms and studies of reliability and validity. J Appl Psychol. 1948;32(3):234–47.

Music for Speech Disorders

Yune Sang Lee, Michelle Wilson, and Kathleen M. Howland

Introduction

There is a growing body of evidence linking speech, language, and music, whereby these faculties have both distinct and shared features. For decades, music has been used to augment or bypass faulty neuroanatomical pathways linked to language/speech dysfunction. In this chapter, we provide a concise review and discussion of the current evidence for music-based interventions and music therapy for speech and language disorders.

Self-Location

Yune Sang Lee, Ph.D. is the director of the Speech Language and Music Lab, and Assistant Professor in the School of Behavioral and Brain Sciences at the University of Texas at Dallas. He received his PhD in cognitive neuroscience from Dartmouth College and postdoctoral training at the University of Pennsylvania. His current research focuses on the behavioral and neural connections between speech,

---

Y. S. Lee (✉)
Department of Speech, Language, and Hearing, School of Behavioral and Brain Sciences, The University of Texas at Dallas, Richardson, TX, USA
e-mail: yune.lee@utdallas.edu

M. Wilson
Department of Otolaryngology—Head and Neck Surgery, Johns Hopkins Healthcare and Surgery Center, Bethesda, MD, USA
e-mail: mwils110@jhmi.edu

K. M. Howland
Department of Music Therapy and Liberal Arts, Berklee College of Music, Boston, MA, USA
e-mail: khowland@berklee.edu

© The Author(s), under exclusive license to Springer Nature Switzerland AG 2023
K. Devlin et al. (eds.), *Music Therapy and Music-Based Interventions in Neurology*, Current Clinical Neurology,
https://doi.org/10.1007/978-3-031-47092-9_6

language, and music with the goal of developing music-based intervention programs for individuals with language and cognitive deficits.

Michelle Wilson, MM, MS, CCC-SLP (she/her) is a speech-language pathologist with Johns Hopkins Department of Otolaryngology—Head and Neck Surgery. She specializes in dysphagia related to head and neck cancer, voice disorders, and gender-affirming voice training. Her clinical practice is heavily informed by her singing training, and she serves as the assistant director for the ParkinSonics™, a Baltimore-based choir for people with Parkinson's Disease.

Kathleen M. Howland, Ph.D. is a music therapist (MT-BC, NMT) and speech-language pathologist (CCC-SLP, LSVT certified). She is a Professor of Music Therapy and Liberal Arts at the Berklee College of Music in Boston. She has worked in neurogenic disorders for 40+ years, specializing in very young children with autism, people who have had strokes and those diagnosed with Parkinson's disease. She has long worked and taught for evidence-based practices based on a biological rationale for interventions targeting nonmusical skills. She has lectured nationally and internationally to disseminate the science of art and the art of science to the benefit of those with diseases and disorders in movement, speech/language, cognition and emotions.

Definitions and Background

Speech is defined as the way sounds are formed between the articulators (i.e. the teeth, lips, tongue, hard palate, and soft palette), the glottis (the voice), and how fluently these parts function together. It is a motor expression of language [1].

Language is a highly representational and symbolic form of communication, conveying linguistic information. Language can be further broken down into *expressive* (generating meaning) and *receptive* (understanding others) language.

Music and speech share many of the same elements, including loudness, pitch, tempo, and rhythm. Pitch, in particular, helps the listener differentiate between a statement and a question. Music and language are "both syntactic systems, employing complex, hierarchically-structured sequences built using implicit structural norms. This organization allows listeners to understand the role of individual words or tones in the context of an unfolding sentence or melody" [2, p. 1]

Auditory processing describes the neural processing and analysis of sounds transmitted by the ear. The ear acts as the microphone, passing along sound from the ear drum, to the inner ear, along the eighth (vestibulocochlear) cranial nerve, and to the brainstem feeding information to the brain. Through this delicate and complex process, the sound waves are converted into electrical and chemical signals interpreted by the brain.

Auditory processing begins in the womb with studies showing fetal reactions to music, including recognition of previously heard music [3–5]. The last trimester in the womb provides the fetus with abundant opportunities to listen to the outside world. This listening facilitates the connections between the recently formed ear anatomy to the brain's auditory cortex.

Studies have demonstrated that there are several elements of music perception present at birth, namely rhythm perception [6], song recall [7], emotional responses to music [8–10], communicative musicality [11], preference for music versus speech [12], pitch discrimination [13–15], preferences for consonance versus dissonance [16], attunement to changes in key [17], and vocalization using musical intervals [18]. Infants as young as 2.5 months vocalize in musical intervals that reflect the way they perceive sound in the world [19]. The degree of vocalization complexity at this age could predict language outcomes at 2 years of age, predicting developmental challenges down the road [20]. Because of these intrinsic music perception abilities, people need not be musicians in order to benefit from music therapy and music-based interventions (MBI).

Parents tend to use infant directed speech (IDS) with babies, which slows the rate of words and uses pitch to highlight particular words in support of language learning [21]. The brains of infants have shown enhanced responsivity to melodically rich streams of information, with degree of response correlated to expressive vocabulary skills at 18 months [22].

Literature Review

Speech and Language Development

Language and music processing both require auditory temporal processing—the active tracking and analyses of rapid and time-varying acoustic signals at multiple hierarchical levels [23, 24]. The brain operates rhythmically, in the form of neural oscillatory entrainment, which is the synchronization of neural oscillations to external auditory stimuli. This allows for complex hierarchical processing and prediction of auditory information, which is vital for real-time acquisition, production, and comprehension of language [25, 26]. Due to its importance to language functions, atypical entrainment of neural oscillations to speech and language is frequently observed in developmental language disorders [27, 28]. Atypical entrainment is in turn correlated with poor phonological awareness and reading skills [29] in these individuals, providing further support for atypical temporal processing as a source of language dysfunction in developmental language disorder. Evidence from neuroimaging and animal neurophysiology highlights the sensorimotor network consisting of cortical (dorsolateral prefrontal cortex, pre-supplementary motor area, supplementary motor area) and subcortical regions (basal ganglia) as common neural resources shared between speech, language, and music [30, 31]. In particular, the basal ganglia are thought to serve as a central hub of temporal processing; damage to the area can lead to various deficits pertaining to learning and execution of temporally organized behaviors. These manifest as deficits in procedural memory [32] and/or sequential pattern learning [33], necessary for both music and language skills.

Associations between language and rhythm processing have been observed in typically developing children, in which children with better rhythmic ability have greater behavioral performance in tasks that tap into core language (e.g., syntax) and language-supporting functions (e.g., auditory short-term memory) [34–36]. Additional research suggests that extensive music training benefits various language skills, including second language pronunciation [37], phoneme processing, and vocabulary knowledge [38]. Based on the emerging evidence, music has been used as a therapeutic means to mitigate language deficits across diverse populations, including autism spectrum disorder (ASD) [39, 40], dyslexia [41], aphasia [42, 43], and developmental language disorder (DLD) [44]. These attempts are largely based upon the premise that music and language share common neural resources, and that musical activities can improve language functioning, not only by the overlap of functions as outlined above, but also promoting the neural plasticity of the shared cortical and sub-cortical systems [45–47].

Developmental language disorders (DLD), is the term we will use for the all-encompassing group of language disorders that develop in childhood (i.e., dyslexia, DLD). DLD impact approximately 7–10% of school-age children [48, 49]. In light of the negative ramifications of this linguistic dysfunction, timely diagnosis and intervention are warranted. Studies provide a promising outlook on the effectiveness of incorporating music-based interventions in language therapy for DLD. Roden et al. [44] reported that listening to acoustically modified classical music for three 30-min sessions per week over the course of 12 weeks led to greater working memory capacity, speech perception, and phoneme discrimination in 40 kindergarten children with DLD. These benefits were not found in either a control group that received no form of intervention, or a group of children that received pedagogical training. While this study did not provide measures of improved language function beyond phonetic processing, these results suggest that even passive listening to acoustically modified music can lead to improved auditory cognitive capabilities. In a feasibility study, Tan and Shoemark [50] provided six bi-weekly 30-min speech therapy sessions targeting morphology and syntax for two children aged 6 with DLD. For each session, individual goals were targeted using either traditional spoken speech, or speech in music. While both children mastered their syntax goals, at the end of the study both children mastered their goals quicker when language was delivered in music than when language was delivered in traditional spoken speech. Alternatively, neither child made significant gains in morphology for either condition, suggesting that the application of music in language intervention may differentially impact certain linguistic domains (i.e., syntax, morphology).

The Connected Biology of Music and Speech/Language

Aniruddh Patel describes the OPERA hypothesis (2012, 2014) as a hypothetical explanation for the connections between music training and sensory processing. Patel based this hypothesis on the fact that music and speech share both sensory and cognitive processing mechanisms in the brain. Because of this, music training and the changes it creates through neuroplasticity, can influence and impact speech processing and production. The OPERA hypothesis is an acronym that includes:

O—the *overlap* of anatomy that serves both music and speech.
P—*precision* reflecting the enhanced demands that music presents to the brain in contrast to speech.
E—the *emotional* valence of music that is a powerfully salient stimulus to the brain.
R—*repetition* which music envelopes and is essential to making changes in neurologic functioning.
A—*attention,* which is described as focused.

Key to Patel's theory (2012, 2014) is the identification of overlapping neural networks. In people who stutter and people with expressive aphasia, one congenital and the other acquired, music processing is preserved when speech is impaired. On the other hand, children with developmental language issues may have specific underlying musical challenges, underscoring network overlap [51]. Patel's hypothesis helps convey why music therapy and music-based interventions can be effective for people with communication disorders. Repetition, necessary for making neural changes, is far more natural in music than in speech. For example, a set of blues lyrics can repeat itself verbatim and be completely acceptable, while speaking the same word or phrase is unnatural and tedious.

Speech and singing are both motor skills that require complex muscle activation and timing coordination. Post-stroke (see Chap. 3), speech can be quite difficult to rehabilitate given this motor complexity. As identified in motor learning, extensive practice is required to optimize functional motor changes [52]. It requires repetition with feedback for errors in order to improve performance. Below is an example of Melodic Intonation Therapy (MIT), the conventional practice for expressive aphasia [53, 54].

Melodic Intonation Therapy

Melodic Intonation Therapy (MIT) has been the primary treatment of expressive aphasia for decades [53–56] (see Chap. 3). In MIT, two notes are used to intone functional phrases with rhythmic cueing of the left hand to stimulate right hemispheric engagement in support of the damaged left. The originators of the protocol have recommended 1.5 h/day, 5 days/week repeating phrases that evolve in complexity from single words to functional phrases. Each session can include dozens upon dozens of repetitions of the same word of phrase. With functional phrases like, "I'm hungry," "I'm thirsty," or "I'm sleepy," the melody remains the same according to the protocol directions. As MIT uses only 2 notes, and the stressed syllable gets the higher note, these phrases will all sound the same (i.e., *I'm HUNgry, I'm THIRsty*). For many people with aphasia, MIT can be experienced as tedious, negatively impacting home practice. A meta-analysis by Bhogal et al. [57] reported recommendations of 2–3 months of intensive aphasia therapy for positive results. This is where music can be applied. This is where music can be applied.

Singing provides the elements that are key to MIT: slowed articulatory speech, syllable lengthening, chunking, and rhythm. Beyond the 2-note structure, conventionally used in MIT, are all the notes of the scale. When assessing people with aphasia, it is typical to ask them to sing a simple overlearned song such as "Happy Birthday" (to give a common example used in the USA). Generally, nonfluent aphasics are able to do this quite well, which is an essential criterion (relatively intact singing and adequate auditory comprehension) for a recommendation of MIT sessions. The ease of singing overlearned songs does not seem to tax the damaged system. Novel, unfamiliar music can be utilized to challenge the brain. People would have to learn the melody, lyrics, rhythms, arrangements of a new song, which taxes multiple brain systems. Increasing the tempo of the music in small increments pressures the motor system into higher and more efficient functioning.

In a study by Halwani et al. [58], the arcuate fasciculus was shown to have increased density in the undamaged right hemisphere following a course of 40 sessions of melodic intonation therapy. This same structure has been shown to be denser in musicians versus non musicians and further in singers versus instrumentalists [58]. Research by Pani et al. [59] has reported that right hemisphere structures predict poststroke speech fluency. If the homologous right hemisphere is to be engaged in support of the damaged left hemisphere, singing may be used to enhance and perhaps accelerate the effects of MIT.

Other Interventions

Other music-based interventions, namely those from Neurologic Music Therapy (NMT® [60]), can add to the variety of approaches to habilitate or rehabilitate speech with meaningful outcomes. The NMT® techniques of Therapeutic Singing (TS®), Vocal Intonation Therapy (VIT®), Musical Speech Stimulation (MUSTIM®) will be described followed by a case study example using MIT and TS®.

TS® can describe singing for self-expression, pleasure, or performance. In this case, it is used to teach singing fundamentals and repertoire for non-musical, clinical gains. TS® can target practicing a novel linguistic task by using unfamiliar but preferred songs that engages the entire system from intention to articulation, requiring attention, working memory and ongoing monitoring of performance. Singing can be used as a warm-up to prime the vocal system for engagement, to facilitate motor timing and coordination, to control breath, to support and challenge speech production by changing the tempo, and to target specific speech sounds for repeated productions. This intervention is used for people with strokes, Parkinson's Disease (PD) (see Chap. 5), chronic obstructive pulmonary disease (COPD), and brain injuries [61–66].

VIT® uses vocal exercises that target laryngeal control of the voice, namely respiration and phonation. Choral warm-ups are the type of activities that are used in VIT®. VIT® exercises can target sustained vocalizations so that a person doesn't run out of breath or volume while communicating; this may be achieved by singing vocal selections with longer sustained phrases, or with louder a dynamic range [61, 65, 67–71].

Case Studies

Case 1

A 60-year-old man (EH) with left side stroke and non-fluent aphasia. He was living in a family home and was a year post-stroke. He was reported to be able to understand what was said to him, but unable to generate reliable responses. His "yes" could be "no," and his "no" could be "yes." The patient had been a corporate manager with excellent linguistic and cognitive faculties prior to his stroke.

On initial assessment, EH was able to sing overlearned songs with ease. In order to engage the plasticity of his brain, we worked extensively with unfamiliar songs. EH was a fan of Tom Petty, whose song catalog was broad. EH's pitch was inaccurate at best, making the point that singing on key, while crucial for auditions to music school, is not crucial for aphasia treatment.

The music therapist created a corpus of over 100 songs, giving EH agency in what he would sing, providing challenges balanced with ease. When he became fatigued, overlearned songs provided a break for him. Those would be well-known Tom Petty songs, as well as those from Neil Diamond and other artists from his CD collection.

Over several months, spontaneous generative speech improved and he was observed using novel, untrained words in a variety of social situations. EH went from single word utterances (*yes, no, wow*) to multi-word phrases learned through music (*I'm getting better all the time*) and multi-word phrases learned through MIT (*I need a break*) with spontaneous utterances related to his occupation, sports and family. to his occupation, sports and family.

Case 2

People with PD may develop reduced speech volume and clarity as the disease progresses due to impairments with muscle activation resulting in bradykinesia and hypokinesia, impacting the respiratory, phonatory, and articulatory systems. Furthermore, PD impacts sensory processing, whereby the individual with PD may be unaware that their speech has become quiet and muffled. Given the degenerative nature of PD, intervention generally targets maintenance where improvements cannot be made or sustained. The ParkinSonics™ was formed in 2015 as part of a randomized, cross-over effectiveness study for people with PD, where participants were placed into two groups – a singing group, and a facilitated discussion group. Each cohort was given 12 weeks of singing or speaking intervention, then crossed over for an additional 12 weeks. Butala et al. [72] found that overall, there was significant improvement in both minimal and average speaking volumes, as well as quality-of-life benefits sustained up to 6 weeks post-intervention. Following the conclusion of the study, participants found they enjoyed the singing group so much, they requested to continue the program; the ParkinSonics™ became a regular singing group based in Baltimore, MD.

The group meets weekly for approximately 35 weeks out of the year for 1.5 h at a time. Rehearsals follow a general agenda (times approximate):

- Dance/movement warm-up (10 min)
- Breathing exercise (10 min)
- Voice exercise (10 min)
- Musical reading (15 min)
- Rhythmic exercises (15 min)
- Practice repertoire (30 min)

The group is run by a professional choir director/pianist, a dance therapist, and a speech-language pathologist/vocalist. The dance therapist runs the movement portion, usually involving body work to move and stretch the limbs and address posture. Breathing exercises focus on thoracic and abdominal engagement, sustained exhalation through pursed lips or fricative sounds, and metronome-based respiratory patterns to facilitate coordination between inhalation and exhalation. These exercises are then carried over to voice exercises utilizing principles of exercise physiology for phonation resistance training [73] as a way to improve vocal power and reduce vocal effort. This is done with a prolonged inhalation to a loud /ja/ on a descending 5-note scale. Participants are encouraged to maintain a louder voice for the duration of the exercise. A recording was also provided to support home practice outside of rehearsals.

Repertoire for the group is chosen based on pitch range, complexity of text and rhythm, and participant interest. Two- and three-part harmony has also been included in repertoire. The education and rhythm exercise portions are based on the selected repertoire.

Conclusion and Future Considerations

In this chapter, we have discussed current research on the application of music-based interventions and music therapy approaches to mitigate language dysfunction in children with developmental language disorders and adults with acquired speech and language disorders. By leveraging music-language connections, researchers have begun to develop and employ effective music-based interventions to successfully improve linguistic function in children with dyslexia and DLD, as well as their typically developing peers. While the enthusiasm surrounding MBI is well deserved, there are several points of caution that must be acknowledged. First, there are very limited measures of language function beyond those of phonological awareness in most, if not all, current studies. Language is a vast cognitive ability, encompassing not only phonology, but also semantics, syntax, morphology, and pragmatics, none of which exist in isolation from one another. While the root cause and profile of each developmental disorder differs, neither dyslexia nor DLD are solely impacted in phonology. Indeed, for music-based interventions to enter into mainstream application for language therapy, a more holistic view of their efficacy and impact on other linguistic areas must be thoroughly investigated. Furthermore, scientific rigor in music therapy research must increase, and there exists an opportunity for greater collaboration with researchers who have a linguistics background. Collaboration is a two-way street, can increase the recognition and esteem of music therapy across neuroscience fields and provide an avenue to translate the work of neuroscience into practical application.

An important point of caution when interpreting studies reporting on language and music-based therapies is a lack of proper control and comparison groups. While certainly not the case in all studies [74, 75] in many instances music intervention for language improvement has been compared to non-linguistic (i.e., art- and dance-based) therapies. Although they provide insight to the effectiveness of alternative interventions, they fail to provide evidence of the benefits of music-based interventions when compared with traditional speech and language therapy. Without a direct comparison to currently employed language interventions, the evidence quality supporting the addition of music-based therapies to these already established treatment protocols remains limited. Employing language therapy as an active comparison group would provide greater support for the inclusion of music and rhythm measures in language therapy.

Finally, a pervasive and well-documented critique of research on developmental language disorders and music-based interventions is the lack of sufficient sample sizes that result in inadequate statistical power for generalizable results. While this is a noteworthy topic, it is important to highlight that this issue of small sample sizes is not limited to research in dyslexia and DLD but permeates research on nearly all clinical populations. Indeed, difficulty in recruitment as well as lack of access to clinical populations make this area particularly difficult to address. Future studies will need to address the issue of small sample sizes to bring music-based interventions into mainstream language therapy. Although challenging, this can be accomplished either through multi-site research and/or by conducting experiments with rigorous planned statistical methods that can appropriately account for the small sample sizes.

We also note a scarcity of neuroimaging research providing brain-based evidence. Owing to advances in neuroimaging techniques including high-resolution and noise-attenuated fast imaging, the future of neuroimaging is well-suited to elucidate detailed neuroanatomical architecture underpinning developmental language disorders and its intimate connection with musical skills.

Utilizing music-based measures in combination with language provides a promising tool to identify language dysfunction earlier in the developmental timescale than is currently possible. This is of utmost importance due to the protracted nature of language development, particularly in the case of developmental language disorders (i.e., dyslexia and DLD). Taken together, there is strong evidence of the relationship between musical ability, particularly rhythmic ability, and language. Music-based interventions present an exciting avenue to aid in the remediation of language and speech disorders.

References

1. Nicolosi L, Harryman E, Kresheck J. Terminology of communication disorders: speech-language-hearing. 5th ed. Lippincott Williams & Wilkins; 2004.
2. Kunert R, Willems RM, Casasanto D, Patel AD, Hagoort P. Music and language syntax interact in Broca's area: an fMRI study. PLoS One. 2015;10(11):e0141069. https://doi.org/10.1371/journal.pone.0141069.
3. Hepper PG. An examination of fetal learning before and after birth. Ir J Psychol. 1991;12(2):95–107. https://doi.org/10.1080/03033910.1991.10557830.
4. Kisilevsky B, Hains SMJ, Jacquet A-Y, Granier-Deferre C, Lecanuet JP. Maturation of fetal response to music. Dev Sci. 2004;7:550–9. https://doi.org/10.1111/j.1467-7687.2004.00379.x.
5. Partanen E, Kujala T, Tervaniemi M, Huotilainen M. Prenatal music exposure induces long-term neural effects. PLoS One. 2013;8(10):1–6. https://doi.org/10.1371/journal.pone.0078946.
6. Winkler I, Háden GP, Ladinig O, Sziller I, Honing H. Newborn infants detect the beat in music. Proc Natl Acad Sci U S A. 2009;106:2468–71. https://doi.org/10.1073/pnas.0809035106.
7. Wilkin PE. A comparison of fetal and newborn responses to music and sound stimuli with and without daily exposure to a specific piece of music. Bull Counc Res Music Educ. 1995;127:163–9.
8. Masataka N. Preference for infant-directed singing in 2-day-old hearing infants of deaf parents. Dev Psychol. 1999;35(4):1001–5. https://doi.org/10.1037//0012-1649.35.4.1001.
9. Nagy E, Cosgrove R, Robertson N, et al. Neonatal musicality: do newborns detect emotions in music? Psychol Stud. 2022;67:501–13. https://doi.org/10.1007/s12646-022-00688-1.
10. Shenfield T, Trehub SE, Nakata T. Maternal singing modulates infant arousal. Psychol Music. 2003;31(4):365–75. https://doi.org/10.1177/03057356030314002.
11. Malloch SN. Mothers and infants and communicative musicality. Music Sci. 1999;3(1_Suppl):29–57. https://doi.org/10.1177/10298649000030S104.
12. Trehub SE, Nakata T. Emotion and music in infancy. Music Sci. 2001;5:37–61.
13. Háden GP, Németh R, Török M, Winkler I. Predictive processing of pitch trends in newborn infants. Brain Res. 2015;1626:14–20. https://doi.org/10.1016/j.brainres.2015.02.048.
14. Stefanics G, Háden GP, Sziller I, Balázs L, Beke A, Winkler I. Newborn infants process pitch intervals. Clin Neurophysiol. 2009;120:304–8.
15. Trehub SE. The developmental origins of musicality. Nat Neurosci. 2003;6:669–73. https://doi.org/10.1038/nn1084.
16. Masataka N. Preference for consonance over dissonance by hearing newborns of deaf parents and of hearing parents. Dev Sci. 2006;9(1):46–50. https://doi.org/10.1111/j.1467-7687.2005.00462.x.
17. Perani D, Saccuman MC, Scifo P, Spada D, Andreolli G, Rovelli R, Baldoli C, Koelsch S. Functional specializations for music processing in the human newborn brain. Proc Natl Acad Sci. 2010;107(10):4758–63.
18. Dobnig D, Stephan S, Wermke P, Wermke K. It all starts with music—musical intervals in neonatal crying. Speech Lang Hear. 2017;21:1–4. https://doi.org/10.1080/2050571X.2017.1368973.
19. Kottman T, Wanner M, Wermke K. Fundamental frequency contour (melody) of infant vocalisations across the first year. Folia Phoniatr Logop. 2022;75:177–87. https://doi.org/10.1159/000528732.
20. Wermke K, Leising D, Stellzig-Eisenhauer A. Relation of melody complexity in infants' cries to language outcome in the second year of life: a longitudinal study. Clin Linguist Phon. 2007;21(11–12):961–73. https://doi.org/10.1080/02699200701659243.
21. Song JY, Demuth K, Morgan J. Effects of the acoustic properties of infant-directed speech on infant word recognition. J Acoust Soc Am. 2010;128(1):389–400. https://doi.org/10.1121/1.3419786.
22. François C, Teixidó M, Takerkart S, Agut T, Bosch L, Rodriguez-Fornells A. Enhanced neonatal brain responses to sung streams predict vocabulary outcomes by age 18 months. Sci Rep. 2017;7(1):12451. https://doi.org/10.1038/s41598-017-12798-2.

23. Ding N, Melloni L, Zhang H, Tian X, Poeppel D. Cortical tracking of hierarchical linguistic structures in connected speech. Nat Neurosci. 2016;19(1):158–64. https://doi.org/10.1038/nn.4186.
24. Poeppel D, Assaneo MF. Speech rhythms and their neural foundations. Nat Rev Neurosci. 2020;21(6):6. https://doi.org/10.1038/s41583-020-0304-4.
25. Goswami U. A neural basis for phonological awareness? An oscillatory temporal-sampling perspective. Curr Dir Psychol Sci. 2018;27(1):56–63. https://doi.org/10.1177/0963721417727520.
26. Ladányi E, Persici V, Fiveash A, Tillmann B, Gordon RL. Is atypical rhythm a risk factor for developmental speech and language disorders? WIREs Cogn Sci. 2020;11(5):e1528. https://doi.org/10.1002/wcs.1528.
27. De Vos A, Vanvooren S, Vanderauwera J, Ghesquière P, Wouters J. Atypical neural synchronization to speech envelope modulations in dyslexia. Brain Lang. 2017;164:106–17. https://doi.org/10.1016/j.bandl.2016.10.002.
28. Goswami U. A neural oscillations perspective on phonological development and phonological processing in developmental dyslexia. Lang Linguist Compass. 2019;13(5):e12328. https://doi.org/10.1111/lnc3.12328.
29. Kraus N. Atypical brain oscillations: a biological basis for dyslexia? Trends Cogn Sci. 2012;16(1):12–3. https://doi.org/10.1016/j.tics.2011.12.001.
30. Kotz S. Differential input of the supplementary motor area to a dedicated temporal processing network: functional and clinical implications. Front Integr Neurosci. 2011;5. https://www.frontiersin.org/articles/10.3389/fnint.2011.00086.
31. Stevens MC, Kiehl KA, Pearlson G, Calhoun VD. Functional neural circuits for mental timekeeping. Hum Brain Mapp. 2007;28(5):394–408. https://doi.org/10.1002/hbm.20285.
32. Ullman MT. Contributions of memory circuits to language: the declarative/procedural model. Cognition. 2004;92(1):231–70. https://doi.org/10.1016/j.cognition.2003.10.008.
33. Goffman L, Gerken L. An alternative to the procedural–declarative memory account of developmental language disorder. J Commun Disord. 2020;83:105946. https://doi.org/10.1016/j.jcomdis.2019.105946.
34. Carr KW, White-Schwoch T, Tierney AT, Strait DL, Kraus N. Beat synchronization predicts neural speech encoding and reading readiness in preschoolers. Proc Natl Acad Sci. 2014;111(40):14559–64. https://doi.org/10.1073/pnas.1406219111.
35. Gordon RL, Shivers CM, Wieland EA, Kotz SA, Yoder PJ, Devin McAuley J. Musical rhythm discrimination explains individual differences in grammar skills in children. Dev Sci. 2015;18(4):635–44. https://doi.org/10.1111/desc.12230.
36. Lee YS, Ahn S, Holt RF, Schellenberg EG. Rhythm and syntax processing in school-age children. Dev Psychol. 2020;56(9):1632–41. https://doi.org/10.1037/dev0000969.
37. Milovanov R, Tervaniemi M. The interplay between musical and linguistic aptitudes: a review. Front Psychol. 2011;2 https://doi.org/10.3389/fpsyg.2011.00321.
38. Linnavalli T, Putkinen V, Lipsanen J, Huotilainen M, Tervaniemi M. Music playschool enhances children's linguistic skills. Sci Rep. 2018;8(1):1. https://doi.org/10.1038/s41598-018-27126-5.
39. Marquez-Garcia AV, Magnuson J, Morris J, Iarocci G, Doesburg S, Moreno S. Music therapy in autism spectrum disorder: a systematic review. Rev J Autism Dev Disord. 2022;9(1):91–107. https://doi.org/10.1007/s40489-021-00246-x.
40. Nazam F, Husain A. Cognitive impairment and rehabilitation of children and adults with autism spectrum disorder. In: Md Ashraf G, Alexiou A, editors. Autism spectrum disorder and Alzheimer's disease: advances in research. Springer Nature; 2021. p. 301–14. https://doi.org/10.1007/978-981-16-4558-7_15.
41. Habibi A, Cahn BR, Damasio A, Damasio H. Neural correlates of accelerated auditory processing in children engaged in music training. Dev Cogn Neurosci. 2016;21:1–14. https://doi.org/10.1016/j.dcn.2016.04.003.
42. Liu Q, Li W, Yin Y, Zhao Z, Yang Y, Zhao Y, Tan Y, Yu J. The effect of music therapy on language recovery in patients with aphasia after stroke: a systematic review and meta-analysis. Neurol Sci. 2022;43(2):863–72. https://doi.org/10.1007/s10072-021-05743-9.

43. Tomaino CM. Effective music therapy techniques in the treatment of nonfluent aphasia. Ann N Y Acad Sci. 2012;1252(1):312–7. https://doi.org/10.1111/j.1749-6632.2012.06451.x.
44. Roden I, Früchtenicht K, Kreutz G, Linderkamp F, Grube D. Auditory stimulation training with technically manipulated musical material in preschool children with specific language impairments: an explorative study. Front Psychol. 2019;10. https://www.frontiersin.org/articles/10.3389/fpsyg.2019.02026.
45. Levitin DJ, Menon V. Musical structure is processed in "language" areas of the brain: a possible role for Brodmann area 47 in temporal coherence. NeuroImage. 2003;20(4):2142–52. https://doi.org/10.1016/j.neuroimage.2003.08.016.
46. Peretz I, Vuvan D, Lagrois M-É, Armony JL. Neural overlap in processing music and speech. Philos Trans R Soc B Biol Sci. 2015;370(1664):20140090. https://doi.org/10.1098/rstb.2014.0090.
47. Sammler D, Koelsch S, Friederici AD. Are left fronto-temporal brain areas a prerequisite for normal music-syntactic processing? Cortex. 2011;47(6):659–73. https://doi.org/10.1016/j.cortex.2010.04.007.
48. Tomblin JB, Records NL, Zhang X. A system for the diagnosis of specific language impairment in kindergarten children. J Speech Lang Hear Res. 1996;39(6):1284–94.
49. Yang L, Li C, Li X, Zhai M, An Q, Zhang Y, et al. Prevalence of developmental dyslexia in primary school children: a systematic review and meta-analysis. Brain Sci. 2022;12(2):240.
50. Tan EYP, Shoemark H. Case study: the feasibility of using song to cue expressive language in children with specific language impairment. Music Ther Perspect. 2017;35(1):63–70. https://doi.org/10.1093/mtp/miv039.
51. Huss M, Verney JP, Fosker T, Mead N, Goswami U. Music, rhythm, rise time perception and developmental dyslexia: perception of musical meter predicts reading and phonology. Cortex. 2011;47(6):674–89.
52. Kitago T, Krakauer JW. Chapter 8—Motor learning principles for neurorehabilitation. In: Barnes MP, Good DC, editors. Handbook of clinical neurology, vol. 110. Elsevier; 2013. p. 93–103. https://doi.org/10.1016/B978-0-444-52901-5.00008-3.
53. Albert ML, Sparks RW, Helm NA. Melodic intonation therapy for aphasia. Arch Neurol. 1973;29:130–1.
54. Sparks RW, Helm NA, Albert ML. Aphasia rehabilitation resulting from melodic intonation therapy. Cortex. 1974;10:303–16. https://doi.org/10.1016/S0010-9452(74)80024-9.
55. Helm-Estabrooks N, Morgan AR, Nicholas M. Melodic intonation therapy. Austin, TX: Pro-Ed., Inc.; 1989.
56. Norton A, Zipse L, Marchina S, Schlaug G. Melodic intonation therapy: shared insights on how it is done and why it might help. Ann N Y Acad Sci. 2009;1169:431–6. https://doi.org/10.1111/j.1749-6632.2009.04859.x.
57. Bhogal SK, Teasell R, Speechley M. Intensity of aphasia therapy, impact on recovery. Stroke. 2003;34(4):987–93. https://doi.org/10.1161/01.STR.0000062343.64383.D0. Epub 2003 Mar 20.
58. Halwani GF, Loui P, Rüber T, Schlaug G. Effects of practice and experience on the arcuate fasciculus: comparing singers, instrumentalists, and non-musicians. Front Psychol. 2011;2:156. https://doi.org/10.3389/fpsyg.2011.00156.
59. Pani E, Zheng X, Wang J, Norton A, Schlaug G. Right hemisphere structures predict poststroke speech fluency. Neurology. 2016;86(17):1574–81. https://doi.org/10.1212/WNL.0000000000002613. Epub 2016 Mar 30.
60. Thaut MH, Hömberg V. Handbook of neurologic music therapy. Oxford University Press; 2014.
61. Baker F, Wigram T, Gold C. The effects of a song-singing programme on the affective speaking intonation of people with traumatic brain injury. Brain Inj. 2005;19:519–28.
62. Bonilha AG, Onofre F, Vieira ML, Prado MY, Martinez JA. Effects of singing classes on pulmonary function and quality of life of COPD patients. Int J Chron Obstruct Pulmon Dis. 2009;4:1–8.

63. Di Benedetto P, Cavazzon M, Mondolo F, Rugiu G, Peratoner A, Biasutti E. Voice and choral singing treatment: a new approach for speech and voice disorders in Parkinson's disease. Eur J Phys Rehabil Med. 2009;45(1):13–9.
64. Elefant C, Baker FA, Lotan M, Lagesen SK, Skeie GO. The effect of group music therapy on mood, speech, and singing in individuals with Parkinson's disease—a feasibility study. J Music Ther. 2012;49:278–302.
65. Haneishi E. Effects of a music therapy voice protocol on speech intelligibility, vocal acoustic measures, and mood of individuals with Parkinson's disease. J Music Ther. 2001;38:273–90.
66. Schlaug G, Marchina S, Norton A. From singing to speaking: why singing may lead to recovery of expressive language function in patients with Broca's aphasia. Music Percept. 2008;25(4):315–23. https://doi.org/10.1525/MP.2008.25.4.315.
67. Apreleva Kolomeytseva AT, Brylev L, Eshghi M, Bottaeva Z, Zhang J, Fachner JC, Street AJ. Home-based music therapy to support bulbar and respiratory functions of persons with early and mid-stage amyotrophic lateral sclerosis-protocol and results from a feasibility study. Brain Sci. 2022;12(4):494. https://doi.org/10.3390/brainsci12040494.
68. Natke U, Donath TM, Kalveram KT. Control of voice fundamental frequency in speaking versus singing. J Acoust Soc Am. 2003;113(3):1587–93. https://doi.org/10.1121/1.1543928.
69. Ozdemir E, Norton A, Schlaug G. Shared and distinct neural correlates of singing and speaking. NeuroImage. 2006;33(2):628–35. https://doi.org/10.1016/j.neuroimage.2006.07.013.
70. Tamplin J. A pilot study into the effect of vocal exercises and singing on dysarthric speech. NeuroRehabilitation. 2008;23(3):207–16.
71. Wiens ME, Reimer MA, Guyn HL. Music therapy as a treatment method for improving respiratory muscle strength in patients with advanced multiple sclerosis: a pilot study. Rehabil Nurs. 1999;24(2):74–80. https://doi.org/10.1002/j.2048-7940.1999.tb01840.x.
72. Butala A, Li K, Swaminathan A, Dunlop S, Salnikova Y, Ficek B, et al. Parkinsonics: a randomized, blinded, cross-over trial of group singing for motor and nonmotor symptoms in idiopathic Parkinson disease. In: Mirabella G, editor. Parkinson's Dis. 2022;2022:1–14.
73. Belsky MA, Shelly S, Rothenberger SD, Ziegler A, Hoffman B, Hapner ER, Gartner-Schmidt JL, Gillespie AI. Phonation resistance training exercises (PhoRTE) with and without expiratory muscle strength training (EMST) for patients with presbyphonia: a noninferiority randomized clinical trial. J Voice. 2023;37(3):398–409. https://doi.org/10.1016/j.jvoice.2021.02.015. Epub 2021 Mar 16.
74. Bhide A, Power A, Goswami U. A rhythmic musical intervention for poor readers: a comparison of efficacy with a letter-based intervention. Mind Brain Educ. 2013;7(2):113–23.
75. Thomson JM, Leong V, Goswami U. Auditory processing interventions and developmental dyslexia: a comparison of phonemic and rhythmic approaches. Read Writ. 2013;26(2):139–61.

Music for Memory Disorders

Hanne Mette Ridder and Concetta Tomaino

> *We have memories for not only the particulars of a song [...] but also the rich associations that keep the melodies alive for us throughout our life. Memories are not actually lost with dementia or with other brain injuries. [1, p. 1]*

Introduction

Some memories are about facts, and some deeply touch us emotionally. Other memories make our bodies carry out sequences of movements; for example, playing a piece of music we know by heart. We may know a person or a piece of music, but for some reason, cannot recall the name, although it is just on the tip of our tongue. At other times we are invaded by memories or involuntary musical imageries (earworms) that are highly distracting and make us lose focus. Music and memory are tightly connected. Music plays a remarkable role in cognition and reminiscence and is a strong trigger for memories. This chapter will explore the role music may play as an important part of therapeutic interventions for persons with memory disorders. We describe *music-based interventions* carried out by health professionals (potentially in collaboration with a music therapist) and *music therapy approaches*, carried out by a credentialed music therapist.

H. M. Ridder (✉)
Department of Communication and Psychology, Aalborg University, Aalborg, Denmark
e-mail: hanne@hum.aau.dk

C. Tomaino
Institute for Music and Neurologic Function, Mount Vernon, NY, USA

Lehman College, City University of New York, Bronx, NY, USA
e-mail: CTomaino@wartburg.org

© The Author(s), under exclusive license to Springer Nature Switzerland AG 2023
K. Devlin et al. (eds.), *Music Therapy and Music-Based Interventions in Neurology*, Current Clinical Neurology,
https://doi.org/10.1007/978-3-031-47092-9_7

Self-Location

Both authors of this chapter have worked with music and memory with clinical populations for decades. Dr. Concetta Tomaino is Executive Director and Co-founder of the Institute for Music and Neurologic Function in New York, USA, which she co-founded in 1995 with neurologist and author Dr. Oliver Sacks. She is also an adjunct professor at Lehman College of the City University of New York. Tomaino's work has been at the intersection of neuroscience and music therapy. Working full time within a clinical setting has allowed her to observe the impact of daily music therapy interventions in sub-acute rehabilitation as well as the behavioral health and quality of life improvements for those in long-term care.

Dr. Hanne Mette Ridder is a professor of music therapy at Aalborg University in Denmark. She has worked as a music therapist with persons with dementia and carried out research in this area since 1995, with the latest studies focusing on how music therapists train professional caregivers in nursing homes in using musical interactions in a person-centered perspective. Ridder's work is based on an integrative understanding, embracing neuroscience and a psychodynamic approach to ecologically oriented practice with interactive music making.

Definitions

Memory

Memory involves storing and retrieving information and is therefore about how and what we learn. Memory is a complex phenomenon associated with several neural systems. To better understand memory, one may distinguish between short-term memory and long-term memory. The latter includes explicit memory (with semantic and episodic memory for facts and events, respectively), and implicit, procedural memory systems.

Memory is formed by neural pathways in the brain involving formation of new synaptic connections and generation of new neurons in key areas such as the hippocampus. Over the course of a lifetime, memory storage becomes more broadly distributed in the brain and therefore earliest memories tend to persist more strongly than recent memories in the context of disease processes that affect new memory formation, like Alzheimer's disease (AD). Long-term memory is, according to neuropsychologist Elkhonon Goldberg [2], an act of recall of previously stored information. Every time memories are activated, they are slightly changed by embedding them into new contexts. This means that memories undergo constant reconstruction and reconfiguration and therefore are no longer quite the same. Autobiographical memories stay with us [3] and represent our sense of *self* and therefore our identity and personhood [4]. They are a mental representation of who we are, allowing for retrieval of both personal semantic information and episodic memories [5].

AIt takes cognitive effort and working memory capacity to recall explicit memories [6]. In contrast, involuntary autobiographical memories summon personal experiences to mind spontaneously and with no direct attempt at retrieval [7, 8]. They differ from voluntary memories by being more specific and emotional, although they may seem less relevant to a given temporal context. They also differ from flashbacks, which are intrusive and force the individual to re-experience traumatic events [9]. Like episodic explicit memories, involuntary autobiographical memories are most often from childhood and young adulthood, representing the so-called "reminiscence bump". They are recalled during an associative process and prompted by cues (for example, the smell of cardamom) and as such without goal-directed cognitive control.

Spontaneous and involuntary memories have a shorter retrieval time and are important to our survival and well-being. Although spontaneous—for example, when in a relaxed state or engaging in routine work—it is possible to systematically manipulate and predict the retrieval of involuntary memories. When there is an overlap between the cues a person is receiving at a given time (smelling cardamom, feeling the heat from a wood stove) and the information stored in that person's memory, the likelihood of retrieving spontaneous memories (sitting in grandma's kitchen) increases [10]. Cues involve the information present at retrieval and may be objects, locations, or sensory impressions such as pictures, smells, or music (see Fig. 7.1).

In common programs for attention and working memory training, tasks that require cognitive control are used. In contrast to such programs, *effortless training* [11] aims to attain an attention-balanced state that involves no effort to control or manipulate stimuli or objects. For example, in nature exposure, flow experiences, and integrative body-mind training, the trainees observe their own awareness and accept the ongoing experience as it is. It is a way of trying not to try, which engages autonomic control in contrast to cognitive control. Research on effortless training is still in its infancy but shows promising results when it comes to training attention and self-control.

A specific type of autobiographical memories are *nostalgic memories*. Nostalgia etymologically relates to homesickness and expresses a longing to return home. It is a bittersweet desire for a period of the past associated with warm feelings towards loved ones, and incorporates personal, meaningful autobiographical memories that are mostly positive [12]. Nostalgia may involve heartwarming memories and/or

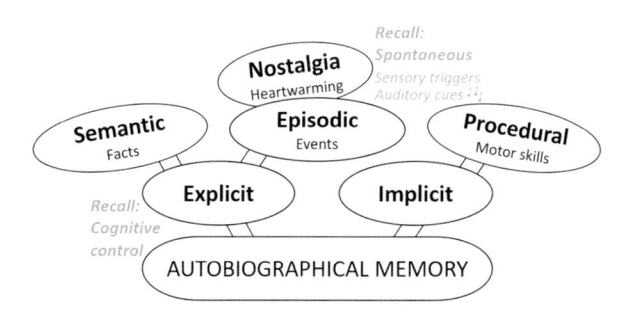

Fig. 7.1 Types of autobiographical memory

wistful grieving for something good but now gone and contributes to making life meaningful and maintaining physiological comfort [13].

Another important type of memory is implicit procedural memory related to motor function. Similar to involuntary memory, procedural memory includes motor skills that, once learned, no longer need to be actively brought to mind. The various motor schema for how to walk, how to dress, and how to speak are mediated at the subcortical level through neural networks involving the cerebellum and basal ganglia [14]. These subcortical networks receive inputs from the auditory system and are influenced by external auditory rhythms. The ability of external *auditory cues* to influence motor function represents a key underlying principle for the use of music therapy in physical rehabilitation programs for people with neurocognitive deficits (see Chaps. 3, 5, and 6).

Memory deficits can be specific and limited, so that a person may have lost memory of recent information but be able to recall rich details about their youth. For persons with AD, working memory and explicit memory are impaired as an early symptom. As the disease progresses, the memory for *semantic* autobiographical information from middle childhood to present life shows a gradient of steady decrease, whereas *episodic* memories have a predominance from age 6 to 30 years, and after this, a steep drop in memories from the age of approximately 30 years [15]. For patients with amnestic mild cognitive impairment, the decline is generally limited to autobiographical memory of recent life [5].

Musical Memory

In his book *Sound and Symbol*, Zuckerkandl explained that memory of an individual tone is not a melody, as it takes a succession of tones to make a melody [16]. The existence of the individual tone in a melody is simultaneously being directed toward what no longer exists and what does not yet exist. Thus; "hearing a melody is hearing, having heard, and being about to hear, all at once." The past is given only as memory, whereas the future is what Zuckerkandl called foreknowledge or forefeeling; within music, past, present and future seems to coexist.

When we recognize a tune, melodic and time relations are mapped onto a stored long-term representation that contains invariant properties of the musical selection [17]. In this understanding, musical memory is a perceptual representation system. Music represents information about the form and structure of events. This means that music is not associated with a fixed semantic system as in speech, although it may convey meaning through other systems, such as emotional analysis and associative memories. It is possible that focal brain damage leads to memory loss specifically limited to music and with selective deficits related to, recognizing or relearning a particular melody. In general, familiar melodies are associated to a series of events related to it, and where *associative* memories contribute to recognition [17].

Melody recognition and associative memories require higher order processing in the brain, involving working memory and temporal and frontal cortex [18]. As musical beat and meter are distinct and processed differently (see Chap. 2), it may lead to working memory overload if there is a lack of synchrony between them. Music processing is a modular process which means that specific processes happen in specific brain areas, in contrast to hierarchical processing, where one process supersedes the other. This means that music tends to promote associative learning when it is recognizable, with clear structures, and decreased sensory demands that do not overload working memory [18].

Musical training regardless of training type can induce brain plasticity, aid memory function, and improve cognitive reserve [19, 20]. Pleasurable experiences such as music listening modulate the brain and can facilitate positive changes in brain plasticity over time [21].

By comparing results from functional magnetic resonance imaging studies in patients with AD and healthy controls, Jacobsen et al. [22] examined why musical memory can be preserved in advanced AD and pointed out an overlap between musical memory regions and areas that are relatively spared in this disease. Based on these findings, they provided a neuroanatomical explanation for the preservation of musical memory in AD and offered evidence for an encoding of musical memory in brain regions that are affected only at late stages of neurodegeneration. Music can thus help persons with AD to preserve their sense of personal identity.

The brain area that seems to be most responsive to multiple aspects of memory associations in autobiographical music is the prefrontal cortex. A 2009 study showed that the prefrontal cortex was engaged when familiar music was associated with several elements including a familiar person, a specific time period and place, and a strong emotional experience [23]. In addition, familiar music has been shown to activate cortical as well as subcortical brain regions (e.g., cerebellum, putamen, limbic system) [24]. In exploring how music-related memory in people with dementia can be assessed comprehensively from a person-centered perspective, Ko and Kim [25] classified musical memory by distinguishing between the role music has as either object or medium. This may be explained as the difference between music *as* memory and music *in* memory. Music *as* memory represents the memories for musical information (e.g., melody, rhythm, and lyric) that can be assessed using tests of melody discrimination, song recognition/familiarity, and musical performance tasks. Music *in* memory stands for the autobiographical memories that music evokes and are connected to the person's identity.

Music as a Memory Trigger

Music is described to have the power to unlock memories and other cognitive capacities in persons with AD [26] and older people generally [27]. Instead of relying on working memory, music may serve as a trigger for spontaneous memories. In this way music is priming or cueing memory retrieval by both external cues and by activating internal states (thoughts and emotions). The most common cues for involuntary, spontaneous memories are sensory impressions, and these unique sounds (e.g., the sudden barking of a dog) tend to predominate in contrast to repeated cues (e.g., bird songs) [10]. Music consists of sounds that unfold over time, as Zuckerkandl described, and the auditory cognitive system thus depends on mechanisms that allow the sound stimulus to relate to a sequence of other sounds that occur later. This involves working memory as processed in the prefrontal cortex. As explained above, music generates associative memories, which serves to strengthen later memory retrieval. When music is matched to mood, it further facilitates memory recall and working memory [28], and when it then further reminds the person of past interactions with friends and family, nostalgic memories are triggered. Among music-induced emotions, nostalgia occupies a prominent position [12]. Notably, there are several pathways in the brain that connect the auditory system and the temporal limbic system through which emotional cues in voice and instrumental music are processed [29]. Thus, in sum, when certain brain areas are damaged by injury or degeneration, there may be alternative subcortical and cortical pathways for evoking musical memories.

Literature Review

Music-Based Interventions

When memories are understood as flexible entities that undergo constant reconstruction and reconfiguration, we must revise how we traditionally think about music biographies and favorite music. If a piece of favorite music is used in rehabilitation of physiological functions for a person in pain, new layers of memories will be added to this piece of music. The painful experiences may change how the music is perceived, and the patient may no longer like what used to be a favorite piece. It is also possible that if a piece of music feels like a helpful enhancer and an emotional support, it may be linked with perceptions of strength and power. In this understanding, the patient may either lose a "good friend" or gain an extra means of support in a specific piece of music. A music therapist may, in this context, help the rehabilitation team and patient in selecting music and facilitating ways music might serve as a support to enhance rehabilitative care. It is important to emphasize that a one-size-fits all approach to music selection is not appropriate in the context of music interventions for memory disorders.

Based on a review of the literature, Matziorinis and Koelsch argue that because AD begins decades before the presentation of clinical symptoms, music interventions might be a promising means to delay and decelerate the neurodegeneration for those who are at risk for developing cognitive decline [33]. Singing in a choir or learning to play an instrument (see Chap. 5) builds up cognitive reserve, which has a protective effect in neurocognitive disorders [30, 31]. Cognitive reserve has, among other features, been associated with expansion of a specific hippocampus area, supporting the importance of cognitive training and for preventive measures in memory disorders [32].

Music Therapy Approaches

In addition to the above, reasons for referring patients with memory disorders to music therapy often include social bonding and existential identity work [33]. Music therapy approaches that specifically address music *in* memory create a setting for sharing meaningful, emotional, personal, and nostalgic memories. Singing or music listening does not necessarily require cognitive control and working memory function. Thus, music may serve as an *effortless* sensory trigger of spontaneous autobiographical memories.

In a study by Dassa, "mutual musicking" (experiencing music together by either playing, singing or listening to music) and concurrently interviewing older persons about the music of their lives strengthened their sense of self-identity. The musicking elicited memories from their youth that emotionally impacted them. Dassa argued that the use of music may support reminiscence, which can promote the following dimensions of quality of life: vitality, agency, social recognition, belonging, and meaning [34].

When it comes to dementia, several meta-analyses show that music therapy significantly improves neuropsychiatric symptoms including depression, anxiety, apathy, aggression, and agitation, while increasing quality of life. However, only one systematic review found an effect of music therapy on *cognitive functions* in persons with dementia [35–39]. In this review four randomized controlled trials were included, but, despite the small number of studies, the results showed a high effect on the improvement of cognitive functions of music therapy carried out by a trained music therapist. For persons with AD, music-evoked autobiographical memories are mostly triggered by songs learned at a young age and are recalled quicker and have more emotional and personal content when the music is self-selected [40].

In a scoping review of tools for assessing music-related memory in people with dementia, Ko and Kim [25] found that out of 47 studies, the majority were about explicit long-term musical memory (semantic, episodic, and autobiographical). They suggest that the assessment of musical memory may provide insights into preserved cognitive functions and that autobiographical memories can help the music therapist understand the client's life. In general, music-evoked nostalgia fosters social connectedness and self-esteem, enhances inspiration, strengthens meaning in life and augments self-continuity [12]. Persons with dementia engage in nostalgic memories in the same way as others with nostalgic autobiographical memories producing more positive affect and increasing expressions of self-esteem, self-continuity, companionship, connectedness, and closeness of relationships. All in all, nostalgic memories reflect life as being meaningful [41]. Importantly, when using music to facilitate prosocial interactions, the music used to trigger memory recall must match the current mood state of the person with dementia [28].

In order to take memory impairments into consideration, we suggest that instead of focusing on explicit semantic memories that require more executive control for voluntary recall (see Fig. 7.1), the music therapist focus on involuntary autobiographical memories that come to mind through processes that require less monitoring. With this approach, the therapist deliberately uses sensory triggers (e.g., songs that evoke spontaneous, nostalgic memories) and in this way, validate the persons' life history and identity.

The use of personalized music should not be limited to treatment of disease symptoms alone but should be integrated as a preventive tool into daily routines at home, at the hospital or in long-term care [42, 43]. Music therapists can play an active role in supporting health professionals in integrating music in daily life and in various ways of effortful or effortless training of memory skills.

If memory deficits make a person feel alienated, anxious, and confused, the person will attempt to communicate their needs for safety to the best of their ability. Sometimes the expressions for these basic human feelings and needs are difficult for others to understand and are often misinterpreted as neuropsychiatric symptoms. In these circumstances, the therapist's deliberate use of musical cues may compensate for loss of memory. Further, the use of music that affords positive recognition and re-cognition may create a safe space for mutual interactions. This framing makes it possible for the music therapist to regulate the person's arousal level by using musical harmony, dynamics, timbre and form to support the person in attaining an attention balanced state. In this state, spontaneous memories come more effortlessly when there are overlaps in cues. Feeling safe and socially connected, and at the same time experiencing music that connects to one's personal identity, may invite nostalgic memories. Sharing such memories in a trusting relationship where emotions are acknowledged and confirmed may be linked to a decrease in neuropsychiatric symptoms as psychosocial needs are met [44].

Practice Example: Community Choirs

Community choirs for people with dementia and their care partners are an example of music-based interventions. In one such program, started as a pilot study in 2011 [45], people with dementia were invited to attend a weekly choral rehearsal with their care partner who also participated in the chorus. A music therapist helped in the design of the program and suggested that a music therapist and a trained choral conductor co-lead the program. The results showed that quality of life and communication with the other members of the dyad improved for those with dementia, and quality of life, social support, communication and self-esteem improved for the caregivers. Most participants stated that benefits included belonging to a group, having a normal activity together and learning new skills [45]. The pilot program was so successful that the participants raised money to sustain the program long term.

Case Example: Music and Autobiographical Memory

Because autobiographical music is so intertwined with an individual's personal history and identity, the actual rendition of a song can be more important than the song itself as in the following example from Concetta Tomaino's work. One resident with memory problems, Mary, is very social and engages in many group programs in the long-term care facility where she resides. She loves music and sings loudly and enthusiastically anytime there is music playing. During a music therapy session (occurring in December) one resident asked for White Christmas, and the music therapist answered by playing a recording of Bing Crosby. Mary started to sing along and kept inserting "Ooh Doop doop, doop doo doop Ooh" after the first few lines but a look of bewilderment came over her. She kept calling out "That's not right" and "It's too slow". The phrase she inserted was something typical in Doo-Wop music of the 1950s. The music therapist looked for a doo-wop version of White Christmas and found a 1958 rendition by The Drifters. After the Bing Crosby version finished, the participants shared memories, and the music therapist then said that there are many versions of this song. As the Drifters version began, Mary immediately shouted "That's it. That's it," after which she freely offered that every Christmas her family got together and sang these songs. Even though a song is familiar, it may not have its fullest impact unless it is the right version of the song. The smile Mary gave the music therapist and the way she leaned back in her chair indicated that this specific version retrieved warm nostalgic memories of social bonding.

Conclusions and Future Considerations

Despite the growing evidence that music therapy and music-based interventions can benefit in the health, wellness, rehabilitation and quality of life of those with memory issues related to neurocognitive deficits including dementia, these services are not widely available - which is notable considering the prevalence of patients with memory disorders. There is enough evidence to demonstrate its efficacy but there are important challenges in music therapy being recognized and reimbursed by healthcare insurance. Music therapy has broad applications and approaches from applying music *as* memory with the neurologic music therapy framework to music *in* memory from an interpersonal, psychotherapeutic approach. The approaches and goals need to be centered on the individual person which impact how some music-based interventions can be generalized as quickly accepted in health and wellness models. There is a movement to increase dialogues across the disciplines of music therapy, neuroscience and community arts programs so that best practices and evidence-based research can further access to these frequently beneficial interventions.

In the future, we would like to see music therapy expanding in two directions. One direction is the integration of music therapy in acute treatment of memory disorders to afford patients feelings of safety and to motivate their training of memory functioning. The other direction is the integration of music therapy in rehabilitation and daily care to provide individual treatment of neuropsychiatric symptoms and to support patients and their caregivers to connect and interact, thereby improving quality of life.

References

1. Tomaino DM. The role of music in the rehabilitation of persons with neurologic diseases: gaining access to 'lost memory' and preserved function through music therapy. MT Today; 2002.
2. Goldberg E. The wisdom paradox. How your mind can grow stronger as your brain grows older. The Free Press; 2005.
3. Strauch B. The secret life of the grown-up brain. Discover the surprising talents of the middle-aged mind. Penguin Books; 2010.
4. Damasio A. The feeling of what happens. Body, emotions and the making of consciousness. Vintage; 2000.
5. Leyhe T, Müller S, Milian M, Eschweiler GW, Saur R. Impairment of episodic and semantic autobiographical memory in patients with mild cognitive impairment and early Alzheimer's disease. Neuropsychologia. 2009;47(12):2464–9.
6. Unsworth N, Spillers GJ, Brewer GA. The role of working memory capacity in autobiographical retrieval: individual differences in strategic search. Memory. 2012;20(2):167–76.
7. Berntsen D. The unbidden past: involuntary autobiographical memories as a basic mode of remembering. Curr Dir Psychol Sci. 2010;19(3):138–42.
8. Floridou GA, Williamson VJ, Emerson LM. Towards a new methodological approach: a novel paradigm for covertly inducing and sampling different forms of spontaneous cognition. Conscious Cogn. 2018;65:126–40.
9. Dominguez E, Casagrande M, Raffone A. Autobiographical memory and mindfulness: a critical review with a systematic search. Mindfulness. 2022;13:1614–51.
10. Berntsen D, Staugaard SR, Sørensen LMT. Why am I remembering this now? Predicting the occurrence of involuntary (spontaneous) episodic memories. J Exp Psychol Gen. 2013;142(2):426.
11. Tang YY, Tang R, Posner MI, Gross JJ. Effortless training of attention and self-control: mechanisms and applications. Trends Cogn Sci. 2022;26(7):567–77.
12. Sedikides C, Leunissen J, Wildschut T. The psychological benefits of music-evoked nostalgia. Psychol Music. 2021;50:1–19. Preprint.
13. Zhou X, Wildschut T, Sedikides C, Chen X, Vingerhoets AJJM. Heartwarming memories: nostalgia maintains physiological comfort. Emotion. 2012;12(4):678–84.
14. Tomaino CM. Pre-learned skills and role of subcortical information processing to maximize rehabilitative outcomes bridging science and music-based interventions. Healthcare. 2022;10(11):2207.
15. Berntsen D, Kirk M, Kopelman MD. Autobiographical memory loss in Alzheimer's disease: the role of the reminiscence bump. Cortex. 2022;150:137–48.
16. Zuckerkandl V. Sound and symbol. Music and the external world. Princeton University Press; 1956.
17. Peretz I, Zatorre RJ. Brain organization for music processing. Annu Rev Psychol. 2005;56(1):89–114.
18. Hernandez-Ruiz E. How is music processed? Tentative answers from cognitive neuroscience. Nord J Music Ther. 2019;28(4):315–32.
19. Romeiser JL, Smith DM, Clouston SA. Musical instrument engagement across the life course and episodic memory in late life: an analysis of 60 years of longitudinal data from the Wisconsin Longitudinal Study. PLoS One. 2021;16(6):1–15.
20. Zhang L, Fu X, Luo D, Xing L, Du Y. Musical experience offsets age-related decline in understanding speech-in-noise: type of training does not matter, working memory is the key. Ear Hear. 2021;42(2):258.
21. Speranza L, Pulcrano S, Perrone-Capano C, di Porzio U, Volpicelli F. Music affects functional brain connectivity and is effective in the treatment of neurological disorders. Rev Neurosci. 2022;33:1–12. Preprint.
22. Jacobsen JH, Stelzer J, Fritz TH, Chételat G, La Joie R, Turner R. Why musical memory can be preserved in advanced Alzheimer's disease. Brain. 2015;138(8):2438–50.

23. Janata P. The neural architecture of music-evoked autobiographical memories. Cereb Cortex. 2009;19(11):2579–94.
24. Thaut MH, Fischer CE, Leggieri M, Vuong V, Churchill NW, Fornazzari LR, Schweizer TA. Neural basis of long-term musical memory in cognitively impaired older persons. Alzheimer Dis Assoc Disord. 2020;34(3):267–71.
25. Ko B, Kim K. Assessing music-related memory in people with dementia: a scoping review. Aging Ment Health. 2022;27:1–11.
26. Clark CN, Warren JD. Music, memory and mechanisms in Alzheimer's disease. Brain. 2015;138(8):2122–5.
27. Engelbrecht R, Bhar S, Ciorciari J. Planting the SEED: a model to describe the functions of music in reminiscence therapy. Complement Ther Clin Pract. 2021;2021(44):101441.
28. Colverson AJ, Trifilio E, Williamson JB. Music, mind, mood, and mingling in Alzheimer's disease and related dementias: a scoping review. J Alzheimers Dis. 2022;86:1–20. Preprint.
29. Fruhholz S, Trost W, Grandjean D. The role of the medial temporal limbic system in processing emotions in voice and music. Prog Neurobiol. 2014;123:1–17.
30. Pentikäinen E, et al. Beneficial effects of choir singing on cognition and well-being of older adults: evidence from a cross-sectional study. PLoS One. 2021;16(2):e0245666.
31. Romeiser JL, Smith DM, Clouston SA. Musical instrument engagement across the life course and episodic memory in late life. PLoS One. 2021;16(6):e0253053.
32. Carapelle E, Mundi C, Cassano T, Avolio C. Interaction between cognitive reserve and biomarkers in Alzheimer disease. Int J Cell Sci. 2020;21(17):6279.
33. Matziorinis AM, Koelsch S. The promise of music therapy for Alzheimer's disease: a review. Ann N Y Acad Sci. 2022;1516(1):11–7.
34. Dassa A. Musical auto-biography interview (MABI) as promoting self-identity and well-being in the elderly through music and reminiscence. Nord J Music Ther. 2018;27(5):419–30.
35. Wong CTT. Effectiveness of music intervention on cognitive function and neuropsychiatric symptoms in the elderly with dementia: a meta-analysis. Front Nurs. 2022;9(2):143–53.
36. Lam H, Li W, Laher I, Wong R. Effects of music therapy on patients with dementia—a systematic review. Geriatrics. 2020;5(4):62.
37. Pedersen SK, Andersen PN, Lugo RG, Andreassen M, Sütterlin S. Effects of music on agitation in dementia: a meta-analysis. Front Psychol. 2017;8:742.
38. van der Steen JT, et al. Music-based therapeutic interventions for people with dementia. Cochrane Database Syst Rev. 2018;5(5):CD003477.
39. Watt JA, Goodarzi Z, Veroniki AA, Nincic V, Khan PA, Ghassemi M, et al. Comparative efficacy of interventions for aggressive and agitated behaviors in dementia: a systematic review and network meta-analysis. Ann Intern Med. 2019;171(9):633–42.
40. Baird A, Brancatisano O, Gelding R, Thompson WF. Characterization of music and photograph evoked autobiographical memories in people with Alzheimer's disease. J Alzheimers Dis. 2018;66(2):693–706.
41. Ismail S, Dodd E, Christopher G, Wildschut T, Sedikides C, Cheston R. The content of nostalgic memories among people living with dementia. Int J Aging Hum Dev. 2022;94(4):436–58.
42. McDermott O, Ridder HMO, Baker FA, Wosch T, Ray K, Stige B. Indirect music therapy practice and skill-sharing in dementia care. J Music Ther. 2018;55(3):255–79.
43. Huang HC, et al. Reminiscence therapy improves cognitive functions and reduces depressive symptoms in elderly people with dementia: a meta-analysis of randomized controlled trials. J Am Med Dir Assoc. 2015;6(12):1087–94.
44. Ridder HMO. Music therapy for people with dementia. In: Jacobsen SL, Pedersen IN, Bonde LO, editors. A comprehensive guide to music therapy. 2nd ed. Jessica Kingsley Publishers; 2019. p. 300–16.
45. Mittelman MS, Papayannopoulou PM. The unforgettables: a chorus for people with dementia with their family members and friends. Int Psychogeriatr. 2018;30(6):779–89.

Music for Neuro-oncological Disorders

Claudia Vinciguerra, Valerio Nardone, and Matthias Holdhoff

Self-Location

Claudia Vinciguerra is a pianist and neurologist who has a keen interest in MBI in neurology and rehabilitation. Dr. Vinciguerra's background as a neurologist allows her to bring a unique perspective on the application of music therapy in the context of neurological conditions.

Valerio Nardone, a radiation oncologist, collaborates with Claudia Vinciguerra and has co-authored the only existing meta-analysis of music therapy in radiotherapy. As a radiation oncologist, Dr. Nardone has expertise in the use of radiotherapy as a mainstream cancer therapy, alongside surgery and chemotherapy. His collaboration with Dr. Vinciguerra in studying music therapy's impact on patients undergoing radiotherapy adds valuable insights to the chapter.

Matthias Holdhoff, a medical oncologist, contributes his expertise in central nervous system cancers to this chapter. With a background in oncology, particularly in treating cancers of the central nervous system, Dr. Holdhoff brings his experience and knowledge of the challenges faced by patients with brain cancer. Additionally, his personal involvement as a musician playing the cello adds a deeper understanding of the role of music in therapeutic interventions.

C. Vinciguerra
Neurology Unit, University Hospital of Salerno, Salerno, Italy

V. Nardone (✉)
Department of Precision Medicine, University of Campania "L. Vanvitelli", Naples, Italy
e-mail: valerio.nardone@unicampania.it

M. Holdhoff
Department of Oncology, The Sidney Kimmel Comprehensive Cancer Center at Johns Hopkins, Johns Hopkins University School of Medicine, Baltimore, MD, USA
e-mail: mholdho1@jhmi.edu

Together, the combined expertise of these authors in neurology, radiation oncology, and medical oncology, as well as their musical backgrounds, provides a comprehensive and multidisciplinary perspective on the use of MBI in brain cancer care.

Background

About 40% of all people will develop cancer during their lifetime [1]. Brain cancers are relatively rare, accounting for about 1.4% of all cancer diagnoses in the United States, but have significant impact on patients' QoL and survival [2]. They can occur at any age, but the incidence is highest in older adults, with the highest rates seen in those over age 65. The impact of having a brain tumor on a person's well-being is significant, but there is still limited understanding of how symptoms develop, interact with each other, and affect health-related QoL [3]. Patients with glioma often experience more than ten symptoms simultaneously, including fatigue, depression, and cognitive deficits, and these symptoms may influence each other [4, 5]. For example, cognitive deficits related to the tumor and therapy may result in a higher mental load when completing tasks, leading to fatigue, and ultimately, social isolation.

The role of music therapy in oncology opens scenarios aimed at addressing problems in the psychosocial, spiritual sphere, as well as disease-related symptoms. In cancer settings, music therapists lead the way in assessments, and they use various individualized interventions in cancer patients and their families, including relevant elements of music in the context of therapeutic relationships, to address prevalent biopsychosocial and spiritual problems, symptoms and needs [6, 7]. Based on the type of patient and their specific symptoms, a music therapist modifies and adjusts the various MBI in order to provide useful strategies for a better symptoms management.

MBI in Patients with Cancer

In the last few years several studies confirmed the effectiveness of MBI [8–10]. Currently there is a wide range of MBI differentiated by whether they are administered by medical-health personnel (music medicine) or by certified music therapists (music therapy). The authors of this chapter hold the belief that MBI administered by music therapists may be more effective than those applied by other personnel, potentially because thy personalized and individual patients are more actively involved through precise and codified therapeutic protocols—but no trials comparative have been conducted to confirm this.

Table 8.1 summarizes some suggested MBI that may be applied in neuro-oncology.

Table 8.1 The different potential MBI that can be used in neuro-oncology

Music-based intervention	Description
Listening to music	Patients listen to pre-recorded or live music
Playing instruments or singing	Patients play an instrument or sing along with a musician or music therapist
Music composition and improvisation	Patients create or improvise music with the help of a music therapist
Music associated with other forms of art	Patients engage in music-based activities that also incorporate other art forms, such as dance, imagery, and reminiscence

Literature Review

To date, there have been few published studies on MBI that are specifically focused on patients with neuro-oncological conditions, including primary and metastatic brain cancers [11–14]. There is, however, a significant and growing body of data on MBI in context of treatment modalities in other cancers that also apply to neuro-oncology. These include surgery, radiation, chemotherapy and palliative care. The following section summarizes the available data as it applies to these treatment modalities.

Surgery

Walworth et al. published one of the first reports of the use of music therapy in patients with brain cancer in 2008 [15]. In this study the authors examined the effects of live music therapy on quality-of-life indicators, medication usage, and length of stay for brain tumor patients undergoing elective surgical procedures. The study suggests that live music therapy may be beneficial in improving QoL indicators for brain tumor patients undergoing surgical procedures.

Jadavji-Mithani et al. reported on another randomized clinical trial (RCT) aimed to assess the effects of listening to major and minor key music on anxiety levels of patients undergoing awake craniotomy [16]. Results showed that patients enjoyed the music and felt more at ease and less anxious before, during, and after the procedure, regardless of the key distinctions. Patients preferred either major key or minor key music but not a combination of both, and those who preferred major key pieces cited tonality while those who selected minor key pieces cited tempo as the primary factor. The study suggests that music interventions can be beneficial for patients undergoing awake surgery, and further investigation of audio interventions in such procedures is recommended.

Kwong et al. performed a mixed method systematic review aimed to evaluate the potential benefit of music intervention in perioperative neurosurgical management [17]. Quantitative findings showed a significant decrease in anxiety following music intervention in four studies and reduced pain perception in three studies. Qualitative analysis provided additional insight into the factors influencing music intervention's effectiveness. Despite the heterogeneity in study characteristics, the review reveals a potential role for MBI in neurosurgery and highlights the importance of considering qualitative evidence in future studies to better characterize its effectiveness.

Kappen et al. have designed an interesting study protocol for a single-center prospective RCT that aims to investigate the effect of perioperative music on delirium incidence and clinical outcomes in adult patients undergoing a craniotomy [18]. The music group listened to the recorded music through headphones before, during, and after surgery until postoperative day 3, while the control group listened to the standard clinical care. The music consisted of different playlists designed from personal preference gathered from questionnaires at baseline. Delirium incidence and related health outcomes such as length of stay, daily function, QoL, and costs will be assessed.

MBI can also be applied in the context of awake surgery. Wu et al. performed an RCT to investigate the effects of music listening on anxiety levels and physiological responses during awake craniotomy [19]. The study found that music intervention significantly reduced anxiety levels, heartbeat rate, systolic pressure, and diastolic pressure over time. The researchers concluded that music listening can be used as a perioperative nursing care intervention to reduce anxiety levels and improve the quality of care for awake craniotomy patients.

Other Investigators have focused on awake surgery performed on musicians. Scerrati et al. reviewed the use of intraoperative musical performances during awake surgery for musicians with brain diseases (including brain cancers) [20]. The authors found that in cases involving tumor resection or surgical treatment for epilepsies involving eloquent areas in musicians, an intraoperative musical performance could allow a more accurate monitoring of complex function. Several other reports have been published in literature for awake surgery in musicians [21–25]. In this context, Drosos et al. reported on the use of awake brain surgery for an autistic patient, that was successfully treated with an awake craniotomy for a grade III astrocytoma while the patient was listening to his favorite music [26]. The patient remained calm and cooperative even during a focal seizure and the surgeons were capable of removing >80% of the brain tumor. Postoperatively, the patient recovered without any deficits.

Finally, it is noteworthy to underline that music therapy approaches can be successfully used also after surgery, in the context of intensive care units. Hansen et al. have investigated the effects of listening to music on self-reported quality of sleep during daytime rest among patients in the intensive care unit (ICU) [27]. The study was conducted between February and April 2016 in two Danish multidisciplinary ICUs and consisted of 37 patients. The intervention group listened to music for 30 min during daytime rest, while the control group rested without music. The Richards-Campbell Sleep Questionnaire was used to measure the subjective quality of sleep. The results showed that listening to music during daytime rest improved subjective sleep quality in patients in the ICU. Furthermore, there were indications that listening to music reduced the subjective experience of noise in some patients. The study suggests that music can be an effective practice for nurses to improve sleep among patients in the ICU.

Radiotherapy

Patients receiving radiation therapy (RT) for brain cancer require thermoplastic masks that are constructed to assist with immobilization, stabilization, and protection of normal tissue. These masks can cause discomfort, and the process of positioning and stabilizing patients can be difficult, particularly for pediatric cancer patients who may fear immobilization. During radiation therapy, the patient must be alone in the treatment room for approximately 5–10 min, which can be distressing. Measures to improve comfort and reduce distress during treatment are essential but, unfortunately, there is limited research on supportive care interventions to help brain cancer patients during their radiation therapy treatment.

The role of music therapy in patients receiving RT is not well established. Only five studies out of 52 in the last Cochrane Review of Bradt dealt with MT in RT cancer patients. Recently, Nardone et al. performed a systematic review on the effectiveness of MBI in cancer patients undergoing radiation treatment who may experience anxiety, stress, fear, depression, and frustration [28].

O'Callaghan et al. [29] invited paediatric patients and families to engage with a music therapist, which increased communication, self-expression, and creativity. The stories of three patients (two of whom had brain cancer) were reported, with benefits such as stress relief, physical and psychosocial improvements, and enhanced clinical communication observed.

Barry et al. [30] conducted a small single-site RCT involving 11 pediatric patients undergoing RT (of whom 5/11 patients had brain cancer). They investigated the potential use of a compact disc created by the children using interactive computer-based music software that was remixed by the music therapist-researcher. The authors investigated the distress perceptions in patients (using the Kidscope questionnaire) [31], parents and RT professionals. They found no differences in outcomes, as well as lack of congruence among groups. 67% of the children in the standard care group resorted to social withdrawal as a coping mechanism, whereas none of the children in the music therapy group relied on social withdrawal. In a study by Nixon et al. [32], 72% of participants reported that listening to music reduced distress related to wearing a thermoplastic mask during radiation therapy (known as "mask anxiety").

Rossetti et al. assessed the effect of MT on anxiety and distress in patients, evaluating the role of a music session with a trained music therapist before simulation computed tomography (CT) [33]. The findings indicated that MT significantly reduced patient anxiety and distress during the clinical procedure.

Finally, Clemens et al. investigated the effect of MT on the physical parameters such as heart rate, blood pressure, respiratory rate, and oxygen saturation, and corresponding QoL measures using quantitative descriptive scales in oncology patients during radiotherapy [34]. The study involved 57 patients (proportion of brain cancer 17.5%) who participated in weekly MT sessions, which significantly improved and patient comfort, reduced pain and respiratory discomfort. The study suggests that establishing MT in the routine clinical setting should be considered to improve patient outcomes.

In conclusion, the use of music therapy for brain cancer patients undergoing radiation therapy is an area that requires further investigation, particularly because only a small percentage of enrolled patients had brain cancer. One unique consideration relates to the nature of radiotherapy, which requires the patient to undergo this treatment alone and immobilized. As a result, it is not feasible to deliver music therapy simultaneously during radiotherapy, but it can be administered before RT takes place.

Chemotherapy and Other Systemic Therapy

The most studied chemotherapy-related side effect in context of MBI is chemotherapy-induced nausea and vomiting (CINV), which is a feared complication of chemotherapy that affects QOL [35]. A systematic review of the literature identified 10 relatively small (total sample size 33–100 patients) RCTs in adult cancer patients that focused on impact of MBI on CINV as the primary outcome [35]. Overall, these studies found MBI may reduce the incidence of anticipatory CINV and reduce the severity of delayed vomiting. Nine RTCs in adults, conducted in patients with breast (5), ovarian (1), esophageal (1) cancer, leukemia (1), and patients receiving high-dose chemotherapy compared standard antiemetic therapy with or without music-based therapies, using pre-recorded, mainly patient-preferred (8) music. All but two of these studies [36, 37] found a reduction of CINV, with only one trial including music therapy as an intervention [37]. This study compared standard symptom management with or without active (e.g., singing and playing music) plus receptive methods in 66 patients undergoing high-dose chemotherapy and autologous stem cell transplantation (HDC plus ASCT), with 33 patients in the intervention group. There was no impact on nausea, QOL, myelosuppression, abdominal symptoms, mucositis or organ function.

Overall, data on MBI on chemotherapy-related side effects are limited as they are derived from relatively small studies involving heterogeneous patient populations and interventions.

MBI in Palliative Care of Cancer Patients

Symptom management is highly relevant in the treatment of patients with cancers of the CNS, and there is a need to optimize integration of palliative interventions (see Chap. 9) in neuro-oncology practice [38, 39]. There is a growing body of data on MBI related to psychological and physical outcomes including anxiety, depression, mood, hope, pain, fatigue and QOL in patients with other types of cancer. A recent Cochrane review by Bradt et al., identified 81 studies that compared MBI plus standard of care interventions with standard of care alone [10]. There were no studies focused solely on patients with neuro-oncological conditions; however, this symptom spectrum is certainly as relevant to neuro-oncology as to any other field in oncology. The survey found that music therapy may result in anxiety reduction (17

RCTs analyzed with a total of 1381 patients), lead to a small to moderate reduction in depression (12 RCTs, 1021 patients), moderately improve mood (4 RCTs, 221 patients), result in a large increase in hope (2 RCTs, 53 patients), lead to moderate to large improvement in pain (12 RCTs, 632 patients), slightly reduce fatigue (10 RCTs, 498 patients), and result in large improvement of QOL. Overall, the degree of certainty of these observations was reported as low to very low in this large systematic review, which is in part due to heterogeneity of the patient populations and interventions studied. Data in children were even more limited as only 3 RCTs were included in the analysis, but they suggest, with low and very low certainty, that MBI may lead to reduction in pain and anxiety [10].

Perspectives on Clinical Practice and Research Opportunities for MBI in Patients with Cancers of the CNS

Although there are limited data on MBI in patients with brain cancers [13, 14, 19, 40], it is perceivable that music interventions may have a different impact on palliating symptoms in cancers within the CNS compared to those that do not involve the brain. Music is processed by the brain, but it is unclear how tumor location, extent of disease and impact of treatment with surgery, radiation and chemotherapy may alter the perception of and impact of music on the patient. MBI have not been found to be harmful to patients, but the ratings of pain experienced from acute painful stimuli were significantly reduced when individuals listened to music they liked, as compared to listening to disliked music or no music at all [41, 42]. Overall, data in cancer patients and cancer-related palliative care indicate more benefit from active music interventions as opposed to receptive music medicine alone.

There are many opportunities to study MBI in patients with brain cancer. The most homogeneous group to study in this context may be patients with newly diagnosed glioblastoma (GB) that overall face a median overall survival of 15–16 months and most of whom undergo initial surgery (biopsy or resection) radiation (most commonly 60 Gy in 30 fractions) in combination with low-dose temozolomide, followed by, if tolerated, 6 months of adjuvant temozolomide (Fig. 8.1) [43]. Virtually all patients with newly diagnosed GB experience significant fatigue, anxiety related to their diagnosis and treatment, as well as other treatment-related side effects for which MBI has previously been explored. While newly diagnosed GB is not the only neuro-oncological cancer that is treated with this or a similar regimen, this patient cohort may serve as a model to systematically test the potential impact of different types of MBI. If different interventions were studied in this patient population, it would be easier to compare the potential impact of these, which may lead to larger, more definitive clinical trials to test and, if promising, optimize MBI for patients with cancers of the CNS.

If resources allow, it may be an option to add MBI in context of current standard clinical practice in the treatment of patients with cancers of the CNS. However, providing evidence of a benefit for specific symptom control and other outcome

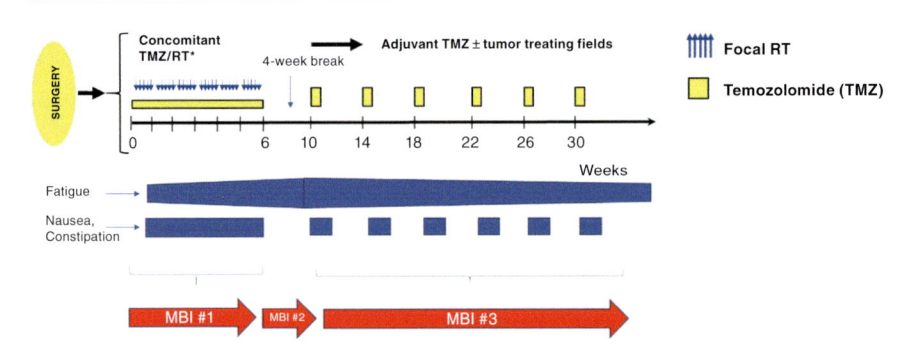

Fig. 8.1 Standard treatment schema for patients with newly diagnosed glioblastoma with suggested incorporation of MBI. Common treatment-related symptoms include fatigue, constipation and nausea. The red arrows indicate distinct phases during treatment that may serve as opportunities to study MBI. While treatment during the first 6 weeks is continuous, there will be a 4-week rest period during which patients are commonly struggling with significant fatigue. For the following 6 months, patients receive high-dose TMZ, which is quite emetogenic, for 5 days, followed by 23 days of rest over 28-day cycles. During these 6 months, days 1–7 of each cycle are commonly the most challenging for patients from a symptom perspective

measures in these cancers may provide a more solid basis to implement this more routinely into clinical practice.

Recently, a meta-analysis of 33 trials that investigated the effects of music interventions combined with standard care compared to standard care alone on anxiety in adult cancer patients [10]. The standardized mean difference of trials that reported post-test anxiety scores on measures other than the full-form State Trait Anxiety Inventory suggested a moderate to large anxiety-reducing effect of music. However, statistical heterogeneity was high across the trials. The text also highlights several limitations of the trials included in the meta-analysis, such as flaws in intervention implementation and high attrition rates, which adversely affects the quality of the evidence [44–53].

Moreover, results of several trials examined the effects of music interventions on depression, distress, and mood in individuals with cancer. The pooled estimate of 12 trials showed a moderate treatment effect of music on depression but with very low certainty evidence, and the results were inconsistent across trials. Sensitivity analyses were conducted to examine the impact of randomization method, the inclusion of participants without cancer diagnosis, and the difference in effect between music therapy and music medicine studies, but these did not significantly impact the pooled effect size. The studies included live music played by music therapists or nursing staff, and participants could not be masked to the music intervention, which may have introduced bias [33, 36, 46–48, 54–58].

Conclusion/Future Considerations

In conclusion, there is a lack of published studies on the effects of MBI specifically tailored to patients with neuro-oncological conditions, including primary and metastatic brain cancers. However, a significant and growing body of data suggests that MBI can be effective in improving the QoL indicators, medication usage, and length of stay for brain tumor patients undergoing surgical procedures. Moreover, MBI has been shown to be effective in reducing anxiety levels and physiological responses during awake craniotomy procedures. The available data suggest that music therapy approaches could be beneficial in enhancing the care of brain cancer patients undergoing surgery. Given the potential benefits of MBI in the context of other cancers and its growing use in other medical settings, it is crucial to conduct further research on MBI for brain cancer patients.

The framework of MBI use in GB patients, as previously stated, could serve as a model to systematically test the potential impact of different MBI.

In the coming decade, there are promising future directions for MBI in the field of neuro-oncology. Building upon the existing research, new avenues of exploration can be pursued to enhance the impact of MBI on brain cancer patients. One area of focus could be the development of tailored music therapy protocols that address specific symptoms and challenges faced by these patients, considering the unique nature of their condition. Furthermore, incorporating technological advancements, such as virtual reality or interactive music platforms, could expand the reach and effectiveness of MBI. Additionally, conducting rigorous clinical trials and longitudinal studies will provide robust evidence for the efficacy of music therapy in improving the QoL for individuals with brain cancer. By collaborating with interdisciplinary teams and integrating neuroscientific approaches, future research can delve deeper into the mechanisms underlying the therapeutic effects of music on cognitive function, emotional well-being, and neuroplasticity in this patient population. Overall, the next decade holds great potential for advancing the field of MBI in neuro-oncology, paving the way for improved support and care for individuals with brain cancer.

References

1. Sung H, Ferlay J, Siegel RL, Laversanne M, Soerjomataram I, Jemal A, Bray F. Global cancer statistics 2020: GLOBOCAN estimates of incidence and mortality worldwide for 36 cancers in 185 countries. CA Cancer J Clin. 2021;71(3):209–49.
2. Forjaz de Lacerda G, Howlader N, Mariotto AB. Differences in cancer survival with relative versus cause-specific approaches: an update using more accurate life tables. Cancer Epidemiol Biomarkers Prev. 2019;28(9):1544–51.
3. Armstrong TS, Vera-Bolanos E, Acquaye AA, Gilbert MR, Ladha H, Mendoza T. The symptom burden of primary brain tumors: evidence for a core set of tumor- and treatment-related symptoms. Neuro-Oncology. 2016;18(2):252–60.
4. Ijzerman-Korevaar M, Snijders TJ, de Graeff A, Teunissen S, de Vos FYF. Prevalence of symptoms in glioma patients throughout the disease trajectory: a systematic review. J Neuro-Oncol. 2018;140(3):485–96.

5. Röttgering JG, Varkevisser T, Gorter M, Belgers V, De Witt Hamer PC, Reijneveld JC, Klein M, Blanken TF, Douw L. Symptom networks in glioma patients: understanding the multidimensionality of symptoms and quality of life. J Cancer Surviv. 2023; https://doi.org/10.1007/s11764-023-01355-8.
6. Magill L. The meaning of the music: the role of music in palliative care music therapy as perceived by bereaved caregivers of advanced cancer patients. Am J Hosp Palliat Care. 2009;26(1):33–9.
7. McClean S, Bunt L, Daykin N. The healing and spiritual properties of music therapy at a cancer care center. J Alternat Complement Med (New York, N.Y.). 2012;18(4):402–7.
8. Bradt J, Dileo C, Grocke D, Magill L. Music interventions for improving psychological and physical outcomes in cancer patients. Cochrane Database Syst Rev. 2011;(8):CD006911.
9. Bradt J, Dileo C, Magill L, Teague A. Music interventions for improving psychological and physical outcomes in cancer patients. Cochrane Database Syst Rev. 2016;(8):CD006911.
10. Bradt J, Dileo C, Myers-Coffman K, Biondo J. Music interventions for improving psychological and physical outcomes in people with cancer. Cochrane Database Syst Rev. 2021;10(10):CD006911.
11. Kappen PR, van den Brink J, Jeekel J, Dirven CMF, Klimek M, Donders-Kamphuis M, Docter-Kerkhof CS, Mooijman SA, Collee E, Nandoe Tewarie RDS, Broekman MLD, Smits M, Vincent A, Satoer D. The effect of musicality on language recovery after awake glioma surgery. Front Hum Neurosci. 2022;16:1028897.
12. Mackel CE, Orrego-Gonzalez EE, Vega RA. Awake craniotomy and intraoperative musical performance for brain tumor surgery: case report and literature review. Brain Tumor Res Treatm. 2023;11(2):145–52.
13. Cheung AT, Li WHC, Ho KY, Lam KKW, Ho LLK, Chiu SY, Chan GCF, Chung JOK. Efficacy of musical training on psychological outcomes and quality of life in Chinese pediatric brain tumor survivors. Psycho-Oncology. 2019;28(1):174–80.
14. Giordano F, Messina R, Riefolo A, Rutigliano C, Perillo T, Grassi M, Santoro N, Signorelli F. Music therapy in children affected by brain tumors. World J Pediatr Surg. 2021;4(3):e000307.
15. Walworth D, Rumana CS, Nguyen J, Jarred J. Effects of live music therapy sessions on quality of life indicators, medications administered and hospital length of stay for patients undergoing elective surgical procedures for brain. J Music Ther. 2008;45(3):349–59.
16. Jadavji-Mithani R, Venkatraghavan L, Bernstein M. Music is beneficial for awake craniotomy patients: a qualitative study. Can J Neurol Sci (Le journal canadien des sciences neurologiques). 2015;42(1):7–16.
17. Ng Kee Kwong KC, Kang CX, Kaliaperumal C. The benefits of perioperative music interventions for patients undergoing neurosurgery: a mixed-methods systematic review. Br J Neurosurg. 2022;36(4):472–82.
18. Kappen P, Jeekel J, Dirven CMF, Klimek M, Kushner SA, Osse RJ, Coesmans M, Poley MJ, Vincent A. Music to prevent deliriUm during neuroSurgerY (MUSYC) clinical trial: a study protocol for a randomised controlled trial. BMJ Open. 2021;11(10):e048270.
19. Wu PY, Huang ML, Lee WP, Wang C, Shih WM. Effects of music listening on anxiety and physiological responses in patients undergoing awake craniotomy. Complement Ther Med. 2017;32:56–60.
20. Scerrati A, Labanti S, Lofrese G, Mongardi L, Cavallo MA, Ricciardi L, De Bonis P. Artists playing music while undergoing brain surgery: a look into the scientific evidence and the social media perspective. Clin Neurol Neurosurg. 2020;196:105911.
21. Piai V, Vos SH, Idelberger R, Gans P, Doorduin J, Ter Laan M. Awake surgery for a violin player: monitoring motor and music performance, a case report. Arch Clin Neuropsychol. 2019;34(1):132–7.
22. Scerrati A, Mongardi L, Cavallo MA, Labanti S, Simioni V, Ricciardi L, De Bonis P. Awake surgery for skills preservation during a sensory area tumor resection in a clarinet player. Acta Neurol Belg. 2021;121(5):1235–9.
23. Bass DI, Shurtleff H, Warner M, Knott D, Poliakov A, Friedman S, Collins MJ, Lopez J, Lockrow JP, Novotny EJ, Ojemann JG, Hauptman JS. Awake mapping of the auditory cortex

during tumor resection in an aspiring musical performer: a case report. Pediatr Neurosurg. 2020;55(6):351–8.
24. Hale MD, Zaman A, Morrall M, Chumas P, Maguire MJ. A novel functional magnetic resonance imaging paradigm for the preoperative assessment of auditory perception in a musician undergoing temporal lobe surgery. World Neurosurg. 2018;111:63–7.
25. Kappen PR, Beshay T, Vincent A, Satoer D, Dirven CMF, Jeckel J, Klimek M. The feasibility and added value of mapping music during awake craniotomy: a systematic review. Eur J Neurosci. 2022;55(2):388–404.
26. Drosos E, Maye H, Youshani AS, Ehsan S, Burnand C, D'Urso PI. Awake brain surgery for autistic patients: is it possible? Surg Neurol Int. 2022;13:543.
27. Hansen IP, Langhorn L, Dreyer P. Effects of music during daytime rest in the intensive care unit. Nurs Crit Care. 2018;23(4):207–13.
28. Nardone V, Vinciguerra C, Correale P, Guida C, Tini P, Reginelli A, Cappabianca S. Music therapy and radiation oncology: state of art and future directions. Complement Ther Clin Pract. 2020;39:101124.
29. O'Callaghan C, Sexton M, Wheeler G. Music therapy as a non-pharmacological anxiolytic for paediatric radiotherapy patients. Australas Radiol. 2007;51(2):159–62.
30. Barry P, O'Callaghan C, Wheeler G, Grocke D. Music therapy CD creation for initial pediatric radiation therapy: a mixed methods analysis. J Music Ther. 2010;47(3):233–63.
31. Buckley L, Curtin M, Cornally N, Harford K, Gibson L. Exploring undergraduate medical student experiences of training within a community-based paediatric clinic. Med Sci Educ. 2023;33(1):73–81.
32. Nixon JL, Brown B, Pigott AE, Turner J, Brown E, Bernard A, Wall LR, Ward EC, Porceddu SV. A prospective examination of mask anxiety during radiotherapy for head and neck cancer and patient perceptions of management strategies. J Med Radiat Sci. 2019;66(3):184–90.
33. Rossetti A, Chadha M, Torres BN, Lee JK, Hylton D, Loewy JV, Harrison LB. The impact of music therapy on anxiety in cancer patients undergoing simulation for radiation therapy. Int J Radiat Oncol Biol Phys. 2017;99(1):103–10.
34. Clemens P, Szeverinski P, Tschann P, Dietl M, Gurk J, Kowatsch M, Tucek G, de Vries A. Physical and nonphysical effects of weekly music therapy intervention on the condition of radiooncology patients. Strahlentherapie und Onkologie: Organ der Deutschen Rontgengesellschaft. 2023;199(3):268–77.
35. Wei TT, Tian X, Zhang FY, Qiang WM, Bai AL. Music interventions for chemotherapy-induced nausea and vomiting: a systematic review and meta-analysis. Support Care Cancer. 2020;28(9):4031–41.
36. Moradian S, Walshe C, Shahidsales S, Ghavam Nasiri MR, Pilling M, Molassiotis A. Nevasic audio program for the prevention of chemotherapy induced nausea and vomiting: a feasibility study using a randomized controlled trial design. Eur J Oncol. 2015;19(3):282–91.
37. Tuinmann G, Preissler P, Böhmer H, Suling A, Bokemeyer C. The effects of music therapy in patients with high-dose chemotherapy and stem cell support: a randomized pilot study. Psycho-Oncology. 2017;26(3):377–84.
38. Walbert T. Integration of palliative care into the neuro-oncology practice: patterns in the United States. Neurooncol Pract. 2014;1(1):3–7.
39. Koekkoek JAF, van der Meer PB, Pace A, Hertler C, Harrison R, Leeper HE, Forst DA, Jalali R, Oliver K, Philip J, Taphoorn MJB, Dirven L, Walbert T. Palliative care and end-of-life care in adults with malignant brain tumors. Neuro-Oncology. 2023;25(3):447–56.
40. Madden JR, Mowry P, Gao D, Cullen PM, Foreman NK. Creative arts therapy improves quality of life for pediatric brain tumor patients receiving outpatient chemotherapy. J Pediatr Oncol Nurs. 2010;27(3):133–45.
41. Martin-Saavedra JS, Vergara-Mendez LD, Talero-Gutiérrez C. Music is an effective intervention for the management of pain: an umbrella review. Complement Ther Clin Pract. 2018;32:103–14.
42. Lu X, Hou X, Zhang L, Li H, Tu Y, Shi H, Hu L. The effect of background liked music on acute pain perception and its neural correlates. Hum Brain Mapp. 2023;44(9):3493–505.

43. Stupp R, Mason WP, van den Bent MJ, Weller M, Fisher B, Taphoorn MJ, Belanger K, Brandes AA, Marosi C, Bogdahn U, Curschmann J, Janzer RC, Ludwin SK, Gorlia T, Allgeier A, Lacombe D, Cairncross JG, Eisenhauer E, Mirimanoff RO. Radiotherapy plus concomitant and adjuvant temozolomide for glioblastoma. N Engl J Med. 2005;352(10):987–96.
44. Alam M, Roongpisuthipong W, Kim NA, Goyal A, Swary JH, Brindise RT, Iyengar S, Pace N, West DP, Polavarapu M, Yoo S. Utility of recorded guided imagery and relaxing music in reducing patient pain and anxiety, and surgeon anxiety, during cutaneous surgical procedures: a single-blinded randomized controlled trial. J Am Acad Dermatol. 2016;75(3):585–9.
45. Bro ML, Johansen C, Vuust P, Enggaard L, Himmelstrup B, Mourits-Andersen T, Brown P, d'Amore F, Andersen EAW, Abildgaard N, Gram J. Effects of live music during chemotherapy in lymphoma patients: a randomized, controlled, multi-center trial. Support Care Cancer. 2019;27(10):3887–96.
46. Cai GR, Li PW, Jiao LP. [Clinical observation of music therapy combined with anti-tumor drugs in treating 116 cases of tumor patients]. Zhongguo Zhong xi yi jie he za zhi Zhongguo Zhongxiyi jiehe zazhi (Chin J Integr Trad Western Med) 2001;21(12):891–4.
47. Chen SC, Yeh ML, Chang HJ, Lin MF. Music, heart rate variability, and symptom clusters: a comparative study. Support Care Cancer. 2020;28(1):351–60.
48. Cassileth BR, Vickers AJ, Magill LA. Music therapy for mood disturbance during hospitalization for autologous stem cell transplantation: a randomized controlled trial. Cancer. 2003;98(12):2723–9.
49. Dóro CA, Neto JZ, Cunha R, Dóro MP. Music therapy improves the mood of patients undergoing hematopoietic stem cells transplantation (controlled randomized study). Support Care Cancer. 2017;25(3):1013–8.
50. Ferrer AJ. The effect of live music on decreasing anxiety in patients undergoing chemotherapy treatment. J Music Ther. 2007;44(3):242–55.
51. Hanser SB, Bauer-Wu S, Kubicek L, Healey M, Manola J, Hernandez M, Bunnell C. Effects of a music therapy intervention on quality of life and distress in women with metastatic breast cancer. J Soc Integr Oncol. 2006;4(3):116–24.
52. Li XM, Yan H, Zhou KN, Dang SN, Wang DL, Zhang YP. Effects of music therapy on pain among female breast cancer patients after radical mastectomy: results from a randomized controlled trial. Breast Cancer Res Treat. 2011;128(2):411–9.
53. Mou Q, Wang X, Xu H, Liu X, Li J. Effects of passive music therapy on anxiety and vital signs in lung cancer patients undergoing peripherally inserted central catheter placement procedure. J Vasc Access. 2020;21(6):875–82.
54. Clark M, Isaacks-Downton G, Wells N, Redlin-Frazier S, Eck C, Hepworth JT, Chakravarthy B. Use of preferred music to reduce emotional distress and symptom activity during radiation therapy. J Music Ther. 2006;43(3):247–65.
55. Arruda MA, Garcia MA, Garcia JB. Evaluation of the effects of music and poetry in oncologic pain relief: a randomized clinical trial. J Palliat Med. 2016;19(9):943–8.
56. Bates D, Bolwell B, Majhail NS, Rybicki L, Yurch M, Abounader D, Kohuth J, Jarancik S, Koniarczyk H, McLellan L, Dabney J, Lawrence C, Gallagher L, Kalaycio M, Sobecks R, Dean R, Hill B, Pohlman B, Hamilton BK, Gerds AT, Jagadeesh D, Liu HD. Music therapy for symptom management after autologous stem cell transplantation: results from a randomized study. Biol Blood Marrow Transplant. 2017;23(9):1567–72.
57. Zhou K, Li X, Li J, Liu M, Dang S, Wang D, Xin X. A clinical randomized controlled trial of music therapy and progressive muscle relaxation training in female breast cancer patients after radical mastectomy: results on depression, anxiety and length of hospital stay. Eur J Oncol Nurs. 2015;19(1):54–9.
58. Ratcliff CG, Prinsloo S, Richardson M, Baynham-Fletcher L, Lee R, Chaoul A, Cohen MZ, de Lima M, Cohen L. Music therapy for patients who have undergone hematopoietic stem cell transplant. Evid Based Complement Alternat Med. 2014;2014:742941.

Music Therapy and Music-Based Interventions for Neurologic Palliative Care

9

Noah Potvin, Maegan Morrow, and Charlotte Pegg

Introduction

Palliative care focuses on the alleviation of suffering over the course of a protracted health journey shaped by an incurable health condition, often prioritizing long-term quality of life (QoL) maintenance. QoL is a complex and subjective construct shaped by individual values, meaning that some patients will prioritize longevity while others will prioritize comfort, with an additional variation regarding the type of preferred comfort (e.g., physical, spiritual, emotional, etc.). Hospice, in contrast, is a philosophy of care introduced further into the disease process when the prognosis is 6 months or less. At this end-of-life stage, QoL goals may shift or focus as patients and caregivers ready themselves for closure and resolution.

As the transition from palliative care to hospice occurs, it is critical that a comprehensive and continuous cultural assessment is undertaken to determine the cultural values informing patients' and caregivers' experiences of the health journey [1]. The integration of cultural values into care promotes a collaborative partnership between the patient, caregivers, and interdisciplinary treatment team that allows for goals and objectives to be collaboratively determined [2]. This partnership also better accommodates shifts in values over the course of the health journey, such as when a patient adjusts their focus from improving gross motor functioning at the time of diagnosis to spiritual or emotional growth closer to the end of life.

N. Potvin (✉)
Duquesne University, Pittsburgh, PA, USA
e-mail: potvinn@duq.edu

M. Morrow
Houston Methodist Hospital, Center for Performing Arts Medicine, Houston, TX, USA
e-mail: mmmorrow@houstonmethodist.org

C. Pegg
Family Pillars Hospice, Bethlehem, PA, USA

© The Author(s), under exclusive license to Springer Nature Switzerland AG 2023
K. Devlin et al. (eds.), *Music Therapy and Music-Based Interventions in Neurology*, Current Clinical Neurology,
https://doi.org/10.1007/978-3-031-47092-9_9

Treatment of neurologic conditions in palliative and hospice care involves a careful alchemy of physical, physiological, and cognitive maintenance in the context of psychological, interpersonal, and spiritual growth as defined by patients' unique cultural locations [3]. This requires for the treatment team to be able and ready to fluidly move among neurobiological, cognitive-behavioral, humanistic, psychospiritual, and existential approaches to care at any given point. In service to that charge, we (the authors) have located our identities and detailed our philosophies below to provide readers a philosophical context for this chapter.

Primarily guiding our integral thinking are culturally reflexive, resource-oriented, and neurologic music therapy philosophies and practices. Culturally reflexive practices focus on how the cultural contexts around the clinician and patient – and the values those contexts impart – impact their respective understandings and expectations of music and health. Relatedly, resource-oriented practices focus on emphasizing intrinsic strengths that promote greater resilience, rejecting illness ideologies that primarily frame health based on deficits [4]. By focusing on strengths, capabilities, and potentialities, resource-oriented practices conceptualize each individual as both whole and a work in progress, thus situating health as an emergent and dynamic construct defined by how one perceives their functioning rather than how that functioning is defined for them.

Neurologic Music Therapy® (NMT ®) is an evidence-based treatment system for board-certified music therapists and other health professionals. In hospice and palliative care settings, NMT® shifts focus to QoL maintenance and away from active rehabilitation of lost functioning. These interventions can be taught to caregivers and family members to address patients' physical, cognitive, and communication goals, including their awareness of their environment and loved ones.

Self-Locations

Noah is a white, non-disabled, straight, agnostic Jewish, cis male living in the regional northeast of the United States. As a board-certified music therapist and licensed professional counselor, he has worked in palliative and hospice care settings for over a decade and practices from a blend of resource-oriented, existentialist, and Gestalt philosophies. Noah's practice is founded on supporting individuals become aware of their intrinsic resources and potentialities; develop understanding of how those resources have been shaped through their cultural ecologies; and learn to activate those resources to develop resilience.

Maegan is a white, neurodivergent, straight, deconstructing protestant Christian cis female living in the southwest region of the United States. Maegan has been practicing music therapy since 2003. Her area of expertise is using NMT® techniques for interdisciplinary work in both the acute and sub-acute (inpatient/outpatient) physical medicine and rehabilitation settings. She has been a fellow in good standing of the Academy of Neurologic Music Therapy since 2006.

Charlotte is a white, non-disabled, queer, agnostic, cis female living in the northeast region of the United States. She works as a board-certified music therapist in the hospice care setting. She practices from a blend of humanistic, resource-oriented, and person-centered philosophies to promote dignity, self-efficacy, and emotional and spiritual resolution as individuals navigate the final chapters of their health journeys.

We are adopting an integral lens through which to explore the role of music therapy and music-based activities in the treatment of individuals with neurodegenerative conditions. Integral thinking avoids attempting to seamlessly integrate two or more inherently incompatible philosophies (e.g., the outcome-oriented work of behaviorism and the process-oriented work of existentialism), and instead explores how disparate approaches can co-exist to provide nuanced care in complex treatment scenario. Through this lens, we negotiate a holistic perspective of health across multiple domains.

Studies with interventions facilitated by interventionists who are board-certified music therapists are often framed as "music medicine" [5]. However, we endorse moving away from this term because it suggests music is an objective science that can be prescriptive or dosed and have chosen to categorize them here as music-based interventions (MBI). Our position is that music is a dynamic humanity that must be individualized to each patient, caregiver, and family system to achieve person-centered and culturally located health outcomes.

Literature Review

There are few studies explicitly exploring music and neurologic disorders in palliative or hospice care; however, since many study participants may have been receiving care primarily focused on QoL, there is the potential for transferability of findings to hospice and palliative care settings. All studies in which the interventionist was a board-certified music therapist (or international equivalent) have been designated as music therapy, and all other studies as MBI.

Across all interventions, it is critical to consider the role of culture in music and health. Traditionally, scholarship exploring music and health has assumed a Western, colonial perspective about the universality of certain music genres and styles; for example, studies examining the impact of pre-recorded selections of classical music prior to a medical procedure. However, there is increased recognition of culture's indelible role in the teaching, modeling, and reinforcing of values that inform how individuals understand and develop relationships with music and health [6, 7].

Music Therapy and Progressive Neurologic Disorders in Hospice and Palliative Care

Music therapy, as an integrative practice at the crossroads of music, psychology, humanities, and neuroscience, is well positioned to be considered a core service in hospice and palliative care in concert within the interdisciplinary treatment team. Music therapy supports whole-person engagement across multiple domains of health that, when addressed as a collective, can facilitate richer health gains [8]. Such an approach is consistent with integrative care, which honors multiple types of health, promotes diverse treatment teams, and cultivates egalitarian collaborations within those teams [9]. In keeping with this philosophy, music therapy – as opposed to its more common designation as "alternative" or "complementary" – can be conceptualized as a core service providing unique health-promoting opportunities of similar value and depth as other treatment team services [10]. When music therapy is positioned as a core service, music therapists can more actively co-treat with team members to provide holistic care responsive to changing goal areas across the disease trajectory and patients gain greater access to those services.

Interventions and Outcomes

Music therapy can provide individuals with neurologic disorders opportunities to cultivate changes in mobility, cognition, expression, interpersonal connectivity, and overall QoL [11–16]. Caregivers also benefit from participation in music therapy as it affords them unique opportunities for direct, and even reciprocal, engagement with their loved ones that are not readily available through verbal mediums [17]. Each of these aspects of health factor into a multidimensional resilience that stands to improve the overall lived experiences of those systems impacted by neurologic illness [18].

Recreative methods involve instrumental and vocal reproduction of pre-composed music to infuse patients' creative expression into music with great meaning to them [19]. The opportunity to determine how a pre-composed work can sound provides control and autonomy often missing due to the progression of their neurologic conditions. Group singing has been shown to exercise cognition, promote mobilization, and improve overall QoL [12, 14, 20], in addition to improving speech, including intensity and phonation [21]. (See Chap. 5 and 6) Instrumental recreation and exercise with music have been found to positively contribute to improved emotional health [22]. Receptive methods – the active, intentional, and mindful listening during the music experience – have been found to help patients process emergent feelings or memories while improving behavioral symptom clusters (e.g., irritability, anxiety, and delusions) [13, 19].

Songwriting – the pre-meditated construction of one or more musical elements to reflect individual expression and identity – has also been recently employed to assist caregivers [19]. Initial scholarship has shown songwriting able to create meaningful therapeutic space for caregivers to explore their caregiving experiences while addressing anxiety, depression, and self-esteem, [17] and to develop deeper relationships with their loved ones based on patient report [16]. Songwriting has been less conclusive: while results showed promise in addressing patients' emotional health, the mixing of songwriting with other interventions confounds the findings [23].

NMT® focuses on two main communication goal areas: (1) comprehension or receiving information (e.g., auditory, visual, and reading comprehension), and (2) communicating information/expressing needs (e.g., verbal, nonverbal and graphic expressions; speech intelligibility; voice quality and volume; and prosody). These communication goals can sustain QoL through interventions (e.g., Music Speech Stimulation®, Melodic Intonation Therapy, Oral Motor and Respiratory Exercises®, Vocal Intonation Therapy®, and Rhythmic Speech Cueing®) that evoke spoken, chanted, and sung vocalizations coupled with specific rhythmic patterns and movements [24–29].

NMT® also focuses on sensorimotor domains. The loss of neurological connections can decrease the ability to control both cognitive processes and coordination of movement, bringing into focus safety considerations related to positioning and movement capabilities. However, rhythmic, harmonic, or melodic cueing through Patterned Sensory Enhancement® and Therapeutic Instrumental Music Performance® facilitates safe active movements that promote interpersonal connections, decrease skin breakdown, prevent pain, and ultimately contribute to QoL maintenance.

MBI for Progressive Neurologic Disorders in Hospice and Palliative Care

MBI in hospice and palliative care can be provided by direct-care staff as a *part* of their practice, whereas for music therapists music *is* their practice. In other words, music in music therapy is the medium for change and transformation, whereas music in MBI amplifies the mediums of change or transformation that other direct-care staff members utilize. While culture is not addressed in these studies, we emphasize once more the importance of cultural expectations of music and music experiences being addressed by the clinician in any music encounter.

MBI can help everyday care be more tolerable for patients by coupling painful or invasive treatments with sensory stimulation from patient-preferred music. MBI also provide caregivers opportunities for experiencing agency in care situations where they might otherwise feel powerless. For example, during the completion of ADLs such as dressing changes or repositioning, direct-care staff can collaborate with caregivers to provide patients receptive music experiences using patient-preferred music to promote emotional regulation, physical relaxation, and interpersonal connection.

Research has explored MBI facilitated by a diverse collection of providers, including musicians, music educators, activities staff and nursing staff [30–34]. Interventions that are facilitated by non-music therapists target similar health goals as music therapists, but by using music as an aspect of their practice, as opposed to the music *being* their practice. MBI have been primarily receptive and recreative with identified health goals including improved mood and overall QoL, eased emotional pain and spiritual distress, improved motor functioning, and enhanced self-efficacy and inclusion [30, 31, 33, 34].

Recreative variations in MBI include drum circles, guitar lessons, and group singing [30–32]. To address feelings of isolation and impeded relationship completion that can come with increased difficulty verbally communicated, [33] facilitators musically trained patients to use music to cultivate new opportunities for communication; this recreation could be adapted (e.g., different instrument selection, repertoire, etc.) in response to declines along the health journey.

Receptive variations require less musical training as recreative variations, making the interventions more accessible to direct-care staff. Pre-recorded music before and during care have been used by nurses to alleviate pain or anxiety and promote life review [33]. Additionally, the act of listening to music in the company of others can simultaneously foster feelings of connectedness with others through health declines and address physiological goals pertaining to relaxation [34, 35].

Intervention and Treatment Outcome Modifications in Hospice and Palliative Care

Patients with neurologic disorders will experience numerous health declines over the course of their health journey including, but not limited to, reductions in strength, mobility, balance, and dexterity; decreased ability to communicate, express, and self-advocate; disrupted executive functioning and short-term memory recall; and increased risk for depression, anxiety, and trauma. These shifts can directly interfere with their ability or readiness to musically engage, and relatedly disrupt an avenue for caregivers to meaningfully connect with them. However, music can be modified to optimize engagement within music encounters, remaining an important medium for addressing physical comfort, interpersonal connectivity, affective range, feelings of agency and autonomy, sensory stimulation, and memory recall. For caregivers in particular, music can afford unique opportunities for direct, and even reciprocal, engagement not readily available through verbal mediums.

Music's capacity to support patients and caregivers throughout the health journey provides an invaluable continuity of care. People form dynamic, evolving relationships with music, and a patient's changing abilities to engage with music through health declines can closely parallel or echo other functional declines in their day-to-day, such as their ability to complete ADLs. Just as engagement with music does not fully cease but instead evolves into new forms of engagement, patients similarly adapt by developing modifications that help them complete desired daily tasks. Music subsequently becomes a medium through which they can explore what it means to gradually lose aspects of independent functioning while, at the same time, amplifying the resources and strengths that the neurologic disorder cannot disrupt.

Psychological safety is a critical consideration through changes in health given how declines that create barriers to musical engagement can result in patient distress. Despite commonly shared notions that music is a safe and non-invasive form of care, music evokes emotions, memories, and experiences that carry the risk of emotional and psychological harm for medically fragile patients who are unable to communicate their distress and may be struggling to access the resources necessary for resilience [36]. This potential for harm sometimes manifests in spaces that are least suspected, such as for patients who identify as musicians but experience a profound shift in self-concept when unable to activate or express their musical identity due to functional decline. *Critically, the elicited distress might be of such severity that patients are contraindicated for music therapy.*

Patient declines can also impact caregivers' emotional and psychological wellness, and consequently their capacity and/or motivation for engaging in music experiences. In other circumstances, the family may wish to discontinue music therapy because music had previously been a shared experience and without the patient's explicit involvement, they are prioritizing other types of care for different outcomes. Subsequently, throughout the health journey it is essential that continuous assessments of patients' physical, physiological, emotional, psychological, and cognitive health are done alongside assessments of caregivers' emotional resilience and psychological readiness. Such assessment provides information for determining how music experiences can remain safe, accessible, and meaningful.

Receptive methods can assume primacy as the patient physically is unable to recreate music or cognitively engage in the process of songwriting. As the patient becomes unable to communicate choice, caregivers can assume a more pronounced role in determining music selection; at the same time, the music therapist relies upon previous assessments to determine music selection. Additional considerations related to sound (e.g., volume and proximity) and musical qualities (e.g., rhythmic intensity, harmonic textures, and lyrical content) help ensure patient safety as they move into fragile medical states wherein they are unable to communicate distress. Receptive NMT® interventions such as Musical Sensory Orientation Training® – the use of live or prerecorded music to stimulate arousal – can be similarly adapted with careful attention to familiar timbres, tones, melodies, rhythms, and life-long songs.

Recreative variations addressing motor functioning are malleable and can address both rehabilitative goals and end-of-life goals that prioritize increased comfort. While interventions centered around movement appear as primarily rehabilitative in nature, they can be adapted in context of hospice and palliative care. For example, recreative interventions targeting increased motor functioning can lead to increased participation in activities of daily living in order to improve mood and/or QoL and to increase feelings of self-efficacy [30].

Case Vignettes

Case 1: A Resource-Oriented Approach

Treatment Context and Goals
Fred was a 56-year-old male with ALS who was receiving hospice care 4 weeks prior to the nurse case manager's referral for music therapy. At the time, Fred was nonspeaking with impaired fine motor and gross motor functionality but sound cognition and pronounced psychological resilience. He readily communicated through typing on a table that streamed directly to a large screen TV in the bedroom. Fred

was interested in storytelling through music therapy to help him prepare for death. We began collaborating on a legacy project involving the development of a slideshow of pictures and accompanying music that Fred selected to represent different periods in his life. Goals for Fred included (a) constructing relevant processes of end-of-life closure, (b) engaging in saying "goodbye" to loved ones, and (c) sustaining psychological resilience through continued physical declines.

Parallel to this process was more psychotherapeutically focused work with his son, Jason. Jason had moved into his adolescence witnessing his once vibrant father transform into a version of himself that he was unable to recognize. Jason's interactions with Fred were overall pleasant and kind but lacked the familiarity and intimacy of a son with his father. Jason appeared to have developed emotional walls around his feelings of loss to protect himself. I encouraged Jason to be a part of this legacy project with his father to so that the dormant father/son relationship could be expressed prior to death. Goals for Jason included (a) establishing and sustaining emotional connectivity with his father, (b) exploring avenues for self-expression related to grief, and (c) exploring ways of communicating "goodbye" to his father.

Treatment Process and Outcomes

With musical and technological assistance, Fred readily engaged in the legacy project. As the project evolved, we had conversations about the songs, artists, and/or genres that represented those categories and I integrated that music into our sessions. Sometimes Fred only wanted to revisit the pre-recorded version and other times he wanted a live recreation to better engage in the here-and-now experience. In those musical encounters, Fred's physical and physiological tension alleviated as his posture shifted, his breathing altered, and his eyes closed.

Throughout this process, Jason remained present but detached. He would offer music to elicit a laugh from his father but did not readily engage in the music recreations nor the storytelling. Fred made frequent requests for Jason to play the guitar accompaniment of John Denver's "Take Me Home, Country Roads," a song he associated with Jason's birth. Jason never outright declined, but he consistently avoided the request.

After 3 months, Fred began imminently dying. Jason agreed to meet that day to record "Take Me Home Country Roads"; while Jason declined to sing because the voice can lead to a vulnerability he was not yet ready for, he played lead guitar while I sang and played rhythm guitar. Afterwards, Jason played the recording for his father, and we realized that Fred's request of the song was him giving Jason permission to say goodbye, and Jason's recording of the song was the authentic goodbye he could provide. Fred died overnight, and his wife reported Jason had been able to lay next to his father and hold his hand in a final act of father and son intimacy.

Case 2: A NMT® Approach

Treatment Context and Goals

Robert was a 65-year-old male with a terminal Glioblastoma. His family requested a music therapist trained in NMT® to provide care in the final months of his life. His daughter, a rehabilitation physical therapist, had observed NMT® interventions helping people with speech issues communicate, and requested individual and family sessions in the home. Robert's wife, three adult children, and two young grandchildren alternated time they were involved in the sessions.

Upon assessment, Robert was confused but without distress or awareness of his declining health. He presented with an expressive aphasia speech disorder due to the glioblastoma's location, leaving him with the difficulty of expressing himself verbally. The focus of care was to provide recreative and improvisational music experiences to support healthy family system interactions and maintenance of Robert's QoL.

Treatment Processes and Outcomes

Through his Jewish identity, music was culturally and spiritually important for Robert. Music making had also been a foundational pastime that he and his children had bonded over throughout the years. Robert and I played guitar and sang familiar, meaningful songs in the initial sessions to develop rapport and assess motor abilities, speech abilities, and guitar skills. When his family became involved, we used Music Speech Stimulation (MUSTIM) ® to retrieve familiar lyrics to preferred songs meaningful to them. Melodic Intonation Therapy (MIT) was also used to stimulate speech from Robert; within this structured intervention, he was able to verbally express 3–4 syllable phrases ("I love you"). These phrases were then recorded in video/audio format as a keepsake to promote positive interaction during his last days.

Through these shared music experiences with his daughters and grandchildren, Robert was able to re-engage with his family. While his young grandchildren were unable to fully grasp Robert's overall decline, they engaged with him in music through laughter, physical gestures, and facial expressions. These creative connections became increasingly important as Robert continued to experience speech decline.

In the last session when Robert was actively declining and sleeping more, I met with him and his adult family members in his bedroom. Live music was played at his bedside while his daughters sang and reminisced about their time with him. Robert passed away later that week in the comfort of his home and with family by his side. It was an honor for me to work with Robert and his family, to provide connection through the powerful medium of music, and to be included in the levaya (memorial service) within his faith community.

Conclusion and Future Considerations

Music therapy and other MBI play an important role in neurologic hospice and palliative care as patients' physical health declines and their emotional and spiritual health – alongside that of caregivers' – assume greater importance. Drawing upon a diverse array of interventions, music therapists can collaborate and co-treat with other interdisciplinary treatment team members to provide substantive and rich holistic care that is culturally informed and responsible. Similarly, direct-care staff implementing MBI stand to expand their practice in ways that allow for additional avenues for patients to improve their QoL and for caregivers to feel supported through pre-bereavement.

Best practices in this area are still in the nascent stages, and there are multiple opportunities for future research-based and practice-based inquiry. Of critical importance is a more explicit focus on the role of cultural values in care. As health declines, values-based medical decision-making becomes more paramount, and different intersections of culturally transmitted values can inform what aspects of care might be prioritized at certain points in the disease process, providing information about the intentions of patients and caregivers within music encounters. Deeper ethnographic perspectives on care can also better inform about how music might hold the potential for harm. For instance, for a patient with historicity in cultures emphasizing the importance of family structures, how can shared music encounters with a caregiver amplify the music's health-promoting effects? All clinicians share similar responsibility to be aware of their culturally located values, expectations, and assumptions about music and health.

An additional area of inquiry includes how interventions can be modified across various stages of decline. Literature has tended to provide cross-sectional perspectives on how music therapy and MBI can be effective in distinct phases of care, but a longitudinal exploration of how music encounters evolve over time has not yet been undertaken. By exploring how patients and caregivers engage with music through substantive shifts in health, clinicians are better positioned to conduct comprehensive assessments.

Finally, future scholarship should address open questions about the most effective clinical processes for utilizing music to help patients maintain agency through their health journey, especially in instances wherein there is dissonance between a patient's cultural values and those of the healthcare system in which they are receiving care. A diverse palette of evidence from qualitative, quantitative, and mixed methods studies could help optimize care for patients and caregivers living through the challenges of progressive neurologic disorders.

References

1. Aronowitz, R. Framing disease: an underappreciated mechanism for the social patterning of health. Soc Sci Med [Internet] [cited 2023 Feb 03]. 2008;67(1):1–9. https://doi.org/10.1016/j.socscimed.2008.02.017. Available from: https://pubmed.ncbi.nlm.nih.gov/18378372/

2. Moore KS, LaGasse AB. Parallels and divergence between neuroscience and humanism: Considerations for the music therapist. Music Ther Perspect [Internet] [cited 2023 Feb 03]. 2018;36(2):144–51. Available from: https://doi.org/10.1093/mtp/miy011.
3. Subramanian I. Integrative medicine in neurology. In: Zigmond MJ, Rowland LP, Coyle JT, editors. Neurobiology of brain disorders [Internet]. London: Academic Press; 2015. p. 953–61. [cited 2023 Feb 03]. Available from: https://ebookcentral.proquest.com/lib/duquesne-ebooks/detail.action?docID=5754459
4. Rolvsjord R. Resource-oriented music therapy in mental health care [Internet]. Gilsum: Barcelona Publishers. 2010. [cited 2023 Feb 03]. Available from: https://search-ebscohost-com.authenticate.library.duq.edu/login.aspx?direct=true&db=cat03556a&AN=duebr.EBC3117628&site=eds-live&scope=site
5. Bradt J, Dileo C, Myers-Coffman K, Biondo J. Music interventions for improving psychological and physical outcomes in people with cancer. Cochrane Database Syst Rev [Internet] [cited 2023 Feb 03]. 2021;10. https://doi.org/10.1002/14651858.CD006911.pub4. Available from: https://pubmed.ncbi.nlm.nih.gov/34637527/
6. Thaut M, Hoemberg V, editors. Handbook of neurologic music therapy. Oxford University Press; 2014. p. 38–325.
7. Lidskog R. The role of music in ethnic identity formation in diaspora: a research review. Int Soc Sci J [Internet] [cited 2023 Feb 03]. 2017;66(219–220):23–38. Available from: https://doi.org/10.1111/issj.12091
8. Hsieh E, Kramer EM. Rethinking culture in health communication: social interactions as intercultural encounters. Hoboken: John Wiley & Sons; 2021. [cited 2023 Feb 03]
9. Porter S, McConnell T, Clarke M, Kirkwood J, Hughes N, Graham-Wisener L, et al. A critical realist evaluation of a music therapy intervention in palliative care. BMC Palliat Care [Internet] [cited 2023 Feb 03]. 2017;16(70). Available from: https://doi.org/10.1186/s12904-017-0253-5
10. Potvin N, Hicks M, Kronk R. Music therapy and nursing cotreatment in integrative hospice and palliative care. J Hosp Palliat Nurs [Internet] [cited 2023 Feb 03]. 2021;23(4):309. https://doi.org/10.1097/NJH.0000000000000747. Available from: https://pubmed.ncbi.nlm.nih.gov/33631776/
11. Devlin K, Alshaikh JT, Pantelyat A. Music therapy and music-based interventions for movement disorders. Curr Neurol Neurosci Rep [Internet] [cited 2023 Feb 03].2019;19(11):83. Available from: https://doi-org/10.1007/s11910-019-1005-0
12. Gómez-Gallego M, Gómez-Gallego JC, Gallego-Mellado M, García-García J. Comparative efficacy of active group music intervention versus group music listening in Alzheimer's disease. Int J Environ Health Res [Internet] [cited 2023 Feb 03]. 2021;18(15):8067. Available from: https://doi.org/10.3390/ijerph18158067
13. Spina E, Barone P, Mosca LL, Forges Davanzati R, Lombardi A, Longo K, Iavarone A, Amboni M. Music therapy for motor and nonmotor symptoms of Parkinson's disease: a prospective, randomized, controlled, single-blinded study. J Am Geriatr Soc [Internet] [cited 2023 Feb 03]. 2016;64(9). Available from: https://doi.org/10.1111/jgs.14295
14. Raglio A, Zaliani A, Baiardi P, Bossi D, Sguazzin C, Capodaglio E, Imbriani C, Gontero G, Imbriani M. Active music therapy approach for stroke patients in the post-acute rehabilitation. Neurol Sci. 2017;38:893–7.
15. Christy JL. A case study of clinical songwriting in music therapy to address emotional expression among an individual with Parkinson's disease and their family caregivers. Doctoral dissertation, Radford University
16. Sharma A, Moon E, Kim G, Kang S. Perspectives of circadian-based music therapy for the pathogenesis and symptomatic treatment of neurodegenerative disorders. Front Hum Neurosci [Internet] [cited 2023 Feb 03]. 2022;15. Available from: https://doi.org/10.3389/fnint.2021.769142
17. García-Valverde E, Badia M, Orgaz MB, Gónzalez-Ingelmo E. The influence of songwriting on QoL of family caregivers of people with dementia: an exploratory study. Nord J Music Ther [Internet] [cited 2023 Feb 03]. 2019;29(1):4–19. Available from: https://doi.org/10.1080/08098131.2019.1630666

18. McCabe P, O'Connor EJ. Why are some people with neurological illness more resilient than others? Psychol Health Med [Internet] [cited 2023 Feb 03]. 2012;17(1):17–34. Available from: https://authenticate.library.duq.edu/login?url=https://search.ebscohost.com/login.aspx?direct=true&db=s3h&AN=69956050&site=eds-live&scope=site
19. Stegemöller EL, Zaman A, Shelley M, Patel B, Kouzi AE, Shirtcliff EA. The effects of group therapeutic singing on cortisol and motor symptoms in persons with Parkinson's disease. Front Hum Neurosci [Internet] [cited 2023 Feb 03]. 2021;15. Available from: https://doi.org/10.3389/fnhum.2021.703382
20. Tamplin J, Baker FA, Grocke D, Brazzale DJ, Pretto JJ, Ruehland WR, et al. Effect of singing on respiratory function, voice, and mood after quadriplegia: a randomized controlled trial. Arch Phys Med [Internet] [cited 2023 Feb 03]. 2012;94(3):426–34. Available from: https://doi.org/10.1016/j.apmr.2012.10.006
21. Pacchetti C, Mancini F, Aglieri R, Fundarò C, Martignoni E, Nappi G. Active music therapy in Parkinson's disease: an integrative method for motor and emotional rehabilitation. Psychosom Med [Internet] [cited 2023 Feb 03]. 2000;62(3):386–393. Available from: http://dx.doi.org.authenticate.library.duq.edu/10.1097/00006842-200005000-00012
22. Kim DS, Park YG, Choi JH, Im SH, Jung KJ, Cha YA, et al. Effects of music therapy on mood in stroke patients. Yonsei Med J [Internet] [cited 2023 Feb 03]. 2011;52(6):977–981. https://doi.org/10.3349/ymj.2011.52.6.977. Available from: https://www.ncbi.nlm.nih.gov/pmc/articles/PMC3220261/
23. Glover H, Kalinowski J, Rastatter M, Stuart A. Effect of instruction to sing on stuttering frequency at normal and fast rates. Percept Mot Skills. 1996;83(2):511–22.
24. Jackson SA, Treharne DA, Boucher J. Rhythm and language in children with moderate learning difficulties. Int J Lang Commun Disord. 1997;32(1):99–108.
25. Thaut MH. The future of music in therapy and medicine. Ann N Y Acad Sci. 2005;1060(1):303–8.
26. Basso A, Capitani E, Vignolo LA. Influence of rehabilitation on language skills in aphasic patients: a controlled study. Arch Neurol. 1979;36(4):190–6.
27. Sparks R, Helm N, Albert M. Aphasia rehabilitation resulting from melodic intonation therapy. Cortex. 1974;10(4):303–16.
28. Haas F, Distenfeld S, Axen K. Effects of perceived musical rhythm on respiratory pattern. J Appl Physiol. 1986;61:1185–91.
29. Bastepe-Gray S, Wainwright L, Lanham DC, Gomez G, Kim JS, Forshee Z, et al. GuitarPD: a randomized pilot study on the impact of nontraditional guitar instruction on functional movement and well-being in Parkinson's disease. Parkinsons Dis [Internet] [cited 2023 Feb 03]. 2022. Available from: https://doi.org/10.1155/2022/1061045
30. Pantelyat A, Syres C, Reichwein S, Willis A. Drum-PD: the use of a drum circle to improve the symptoms and signs of Parkinson's disease (PD). Mov Disord Clin Pract [Internet] [cited 2023 Feb 03]. 2015;3(3):243–9. Available from: https://doi.org/10.1002/mdc3.12269
31. Butala A, Li K, Swaminathan A, Dunlop S, Salnikova Y, Ficek B, et al. Parkinsonics: a randomized, blinded, cross-over trial of group singing for motor and nonmotor symptoms in idiopathic Parkinson disease. Parkinsons Dis [Internet] [cited 2023 Feb 03]. 2022. Available from: https://doi.org/10.1155/2022/4233203
32. Black BP, Penrose-Thompson P. Music as a therapeutic resource in end-of-life care. J Hosp Palliat Nurs [Internet] [cited 2023 Feb 03]. 2012;14(2):118–25. Available from: https://doi.org/10.1097/njh.0b013e31824765a2
33. Garabedian CE, Kelly F. Haven: sharing receptive music listening to foster connections and wellbeing for people with dementia who are nearing the end of life, and those who care for them. Dementia Int J Soc Res Methodol [Internet] [cited 2023 Feb 03]. 2020;19(5):1657–1671. Available from: https://doi.org/10.1177/1471301218804728
34. Zhou Z, Zhou R, Wei W, Luan R, Li K. Effects of music-based movement therapy on motor function, balance, gait, mental health, and QoL for patients with Parkinson's disease: A systematic review and meta-analysis. Clin Rehabil [Internet] [cited 2023 Feb 03]. 2021;35(7):937–51. Available from: https://doi.org/10.1177/0269215521990526

35. Halstead MT, Roscoe ST. Restoring the spirit at the end of life: music as an intervention for oncology nurses. Clin J Oncol Nurs [Internet] [cited 2023 Feb 03]. 2002;6(6):332–6. Available from: https://doi.org/10.1188/02.cjon.332-336
36. Murakami B. The Music Therapy and Harm Model (MTHM): conceptualizing harm within music therapy practice. ECOS. 2021;6

Music for Autoimmune Neurological Disorders

10

Cindybet Pérez-Martínez, Flor del Cielo Hernández, and Jamie Shegogue

Introduction

This chapter aims to define autoimmune neurological disorders, identify the correlating multidimensional needs of the patients living with autoimmune neurological disorders, and provide examples of music therapy interventions that may be used with this to meet those needs. Our research, personal and clinical practice experiences inform this chapter.

Self-Location

Cindybet has been a clinical and neurological music therapist (NMT) certified by the Certification Board of Music Therapists (MT-BC, NMT) for over a decade. I identify as an able-bodied, cisgender, Latina, Puerto Rican female. From early in my career, I have worked with autoimmune neurological patients, particularly those with multiple sclerosis. Having worked with patients in the continental U.S. and in the U.S. territory of Puerto Rico has provided me with a cross-cultural perspective on the perception and treatment of these disorders. My training in visualization and music, neurologic music therapy, and body-mind medicine have given me the tools to explore the biopsychosocial aspects of music therapy treatment in a holistic and

---

C. Pérez-Martínez (✉)
Washington Adventist University, Takoma Park, MD, USA
e-mail: cperez@wau.edu

F. del Cielo Hernández
Florida State University, Tallahassee, FL, USA
e-mail: fhernandez2@fsu.edu

J. Shegogue
Shenandoah University, Winchester, VA, USA

© The Author(s), under exclusive license to Springer Nature Switzerland AG 2023
K. Devlin et al. (eds.), *Music Therapy and Music-Based Interventions in Neurology*, Current Clinical Neurology,
https://doi.org/10.1007/978-3-031-47092-9_10

educational approach and witness the positive results of addressing the client as a whole. In addition, my experiences working with people with diverse neurological disorders including stroke patients, neurointensive care patients, multiple sclerosis and Parkinson's disease, among others, in medical settings and private practice in Puerto Rico and the U.S. are used to create context when writing this chapter.

Flor del Cielo has been a certified music therapist (MT-BC) practicing for over 15 years in the Southeastern region of the United States and identifies as an able-bodied, cisgender, Latina, Puerto Rican female. In addition to working with patients transferred from the neurological ward due to their psychiatric symptoms, I have been a caregiver for loved ones diagnosed with neurological and autoimmune disorders. I have learned about the financial, time investment, and planning associated with caring for a neurological or autoimmune disease patient. My personal experiences became an incentive for my clinical practice, where I used holistic, transdiagnostic, patient-centered approaches, adapting to meet the patient's needs. Both authors were born, raised, and completed their undergraduate degrees in the U.S. territory of Puerto Rico and furthered their music therapy education in the continental U.S.

Jamie is a music therapy student at Shenandoah University in Winchester, Virginia who is completing her clinical internship at the Johns Hopkins Center for Music and Medicine in Baltimore, Maryland. She is a white, cisgender woman with music therapy experience in a variety of patient communities including hospice care, outpatient physical therapy clinics, juvenile detention, and inpatient pediatric units. In this chapter, she has provided student insight on the topics reviewed, offering the perspective of a budding music therapist on the information included.

Definition/Background

Autoimmune disorders are diseases characterized by a dysfunction of the immune system; components of the immune system attack pathogens that enter the body and healthy tissues and organs, resulting in an overall dysfunction of bodily functions [1]. When the areas attacked by the immune system are the central nervous system (brain, spinal cord) and the peripheral nervous system (which includes the peripheral nerves, skeletal muscles, and the neuromuscular junction), the autoimmune disease is considered an autoimmune neurological disease or disorder [2]. The medical community has recognized autoimmune peripheral nervous system diseases since the 1970s [3]. However, autoimmune central nervous system disorders such as autoimmune encephalitis have been more widely studied since 2004 [3].

The grouping under autoimmune neurological disorders encompasses over 40 diagnoses that include multiple sclerosis, encephalopathies, and other central nervous system degenerative conditions [4]. People may experience symptoms affecting their quality of life but may have to wait months or even years to receive an accurate diagnosis and treatment plan [5]. Disease expression may vary from person to person, which may present a challenge when isolating the specific risk factors that may trigger an autoimmune response. Allergic reactions, genetic traits, predispositions, and environmental factors may trigger autoimmune neurological disorders [6, 7].

There is much to discover about autoimmune neurological disorders and their possible causes. There is a lack of long-term studies based on gender, ethnicity, or origin that could truly reveal the factors predisposing people or specific groups [6]. It has been suggested that upper respiratory infections may be one of the main triggers of the chronic inflammatory neurological response that characterizes autoimmune neurological disorders [4]. Other studies have suggested that, at times, the immune system creates antibodies in great numbers without apparent triggers, creating a dysfunctional immune response that negatively impacts the individual's health as the body attacks itself [4, 8].

Most autoimmune neurological disorders have no cure and require chronic symptom management [9]. After a diagnosis is made, early identification of needs and early medical, pharmacological, and nonpharmacological interventions may provide patients with a better quality of life, better management of symptoms, and the skills needed to cope with what could be a significant and permanent change to life as they know it [10].

Symptoms

As can be seen in Fig. 10.1, some of the most common symptoms can be grouped into three domains: physical/medical, cognitive, and psychosocial. Common symptoms were gathered from the National Institute of Neurological Disorders and Stroke and the Genetic and Rare Diseases Information Center web pages.

In autoimmune disorders affecting the brain, psychiatric symptoms can sometimes appear before physical and neurological symptoms [11]. Some diagnoses,

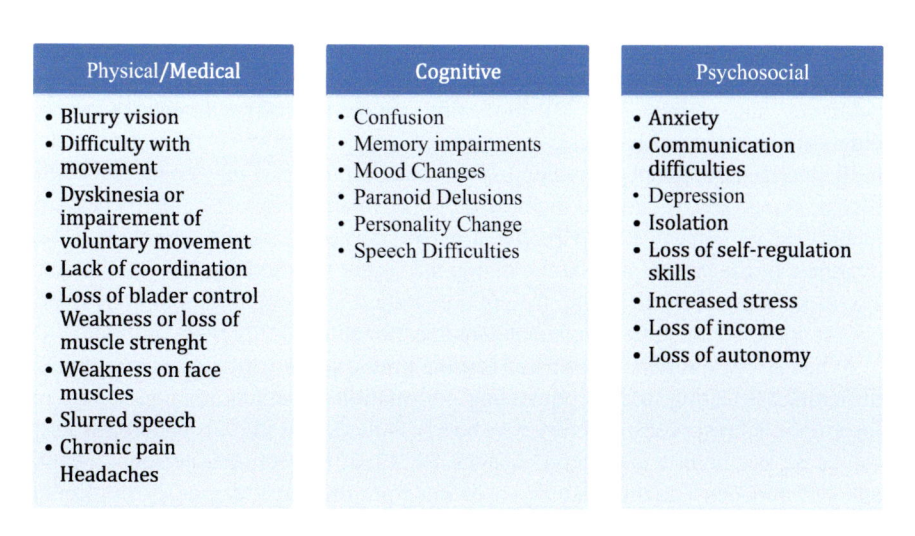

Physical/Medical	Cognitive	Psychosocial
• Blurry vision • Difficulty with movement • Dyskinesia or impairement of voluntary movement • Lack of coordination • Loss of blader control Weakness or loss of muscle strenght • Weakness on face muscles • Slurred speech • Chronic pain Headaches	• Confusion • Memory impairments • Mood Changes • Paranoid Delusions • Personality Change • Speech Difficulties	• Anxiety • Communication difficulties • Depression • Isolation • Loss of self-regulation skills • Increased stress • Loss of income • Loss of autonomy

Fig. 10.1 Frequent autoimmune neurologic symptoms by domain

such as schizophrenia, autism spectrum disorder, depression, and bipolar disorder, may have characteristics in common with autoimmune neurological diseases [12, 13]. As a result, disorders like autoimmune encephalitis have been erroneously diagnosed as psychiatric illness and therefore gone untreated from a neurological medical standpoint [3]. Thus, researchers are considering how chronic inflammation in autoimmune neurological disorders may also affect cognition and impair a person's behavior [14]. Unfortunately, a misdiagnosis or partially correct diagnosis may have significant consequences [15].

Assessments

The physical symptoms of neurological autoimmune disorders directly impact a person's psychosocial well-being and need to by systematically assessed; other assessments are also necessary to address a person's quality of life, mood, and resiliency [16, 17]. For example, frequently used assessments such as ranking scales, Assessment of Quality of Life (AQoL), Visual Analogue Scales (VAS), and the DSM-V Cultural Formulation Interview [18], could become helpful tools for music therapists to understand a patient's perspectives, strengths, challenges, perceptions of the problem and wishes for the future.

Testing and medical procedures may increase a patient's stress level, leading to medical trauma and dysregulation [19–21]. Music therapy sessions have aided in reducing levels of hormones and neurotransmitters associated with stress, which negatively impact the functioning of the immune system [22–24]. Patients participating in music therapy interventions while undergoing medical procedures had lower salivary cortisol levels during these stressful events [25].

A clinician may become aware of nonmedical concerns when assessing for stressors that affect the patient and their support system beyond those related to their medical symptoms [16, 26]. Patients may report financial concerns, having to adapt to loss of familiar social routines, professional skills, coping strategies, and autonomy and identity [26, 27]. Expo a patient's intersectionalities beyond diagnosis, symptoms, and possible treatments may help a music therapist develop individualized treatment, including psychosocial aspects that may connect with the illness, symptoms, social and medical supports, financial aspects, and the person's identity in its entirety [28]. Because a person's intersectionalities provide a rich overview of the person as an entire being, the authors created Fig. 10.2 to illustrate key aspects that will affect the patient's treatment and prognosis while providing opportunities to integrate the patient's intersectionalities.

When an autoimmune disorder affects the limbic system, the person experiences difficulties detecting threats, processing information, accessing memories, providing emotional responses, and restoring bodily homeostasis [4, 29]. The illness itself can cause personality and mood changes, [4, 12, 13] affecting a person's identity and self-concept. Patients may reject or integrate their new diagnosis and limitations as part of their self-concept and personality [30].

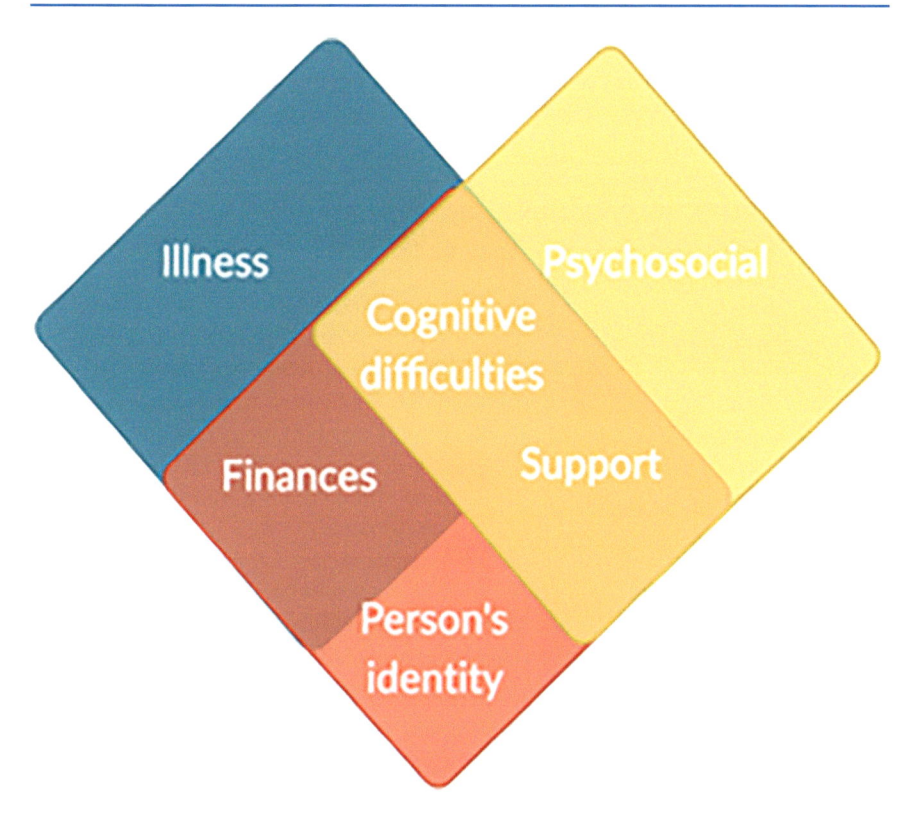

Fig. 10.2 A patient's intersectionalities

Literature Review

Music Therapy and Autoimmune Neurological Disorders

Music therapy is an evidence-informed discipline that can address the multi-dimensionality of humanness and the needs that arise from living with an autoimmune neurological disorder and its treatment [31]. Reports on music therapy interventions when working with multiple sclerosis patients have focused on goals such as decreasing isolation and depressive symptoms, and anxiety; developing coping strategies, and increasing expression, communication skills, and respiratory strength [32–36]; and developing a healthy self-concept and self-esteem [37]. Additionally, active music therapy interventions such as singing, songwriting, and musical instrument playing and receptive (such as music listening), and integrative (such as dancing, movement to music, and drawing) approaches may be used. Researchers have reported that patients with neurological disorders receiving music therapy may regain or maintain personal independence and rehabilitate symptoms such as bradykinesia [38]; further, music therapy may assist in improving synchronized movements [39]. Beyond physical improvements, music therapy interventions can assist patients with neurological conditions in elevating their mood and decreasing their perceived depression levels [40], increasing patients' motivation, and improving aspects of cognitive functioning such as speech production and memory [41].

Music Therapy Approaches and Interventions

The field of music therapy offers diverse approaches and interventions that are effective when supporting patients in achieving social, emotional, and physical goals [42]. Neurologic Music Therapy (NMT®) approaches have been used with neurological autoimmune diseases through the application of techniques such as rhythmic auditory stimulation (RAS®) for gait rehabilitation, patterned sensory enhancement (PSE®) for improvement of physical strength and other motor skills, and therapeutic singing improve voice quality, speech pacing, and volume control [38] (See Table 10.1).

The biopsychosocial approaches and music therapy based on body-mind-medicine may provide clients with music therapy interventions and experiences to stimulate a psychoneuroimmunoendocrinological body response [43–45]. In congruence with a patient-centered model and while considering patient intersectionalities, a psychosocial approach considers social factors (culture, spiritual beliefs, family, and community) in a way that may enhance the quality of life of the clients suffering from autoimmune neurological disorders [22, 46].

The music therapy psychoeducational approach is rooted in the biopsychosocial model and facilitates illness management while including the caregiver in the process [47]. A psychoeducational approach helps organize therapeutic experiences while encouraging patients to process their diagnosis, symptoms, and coping strategies [48] and finding support from others in a group setting [49]. This approach provides a structure for patients and their families to learn and explore their treatment options, rethink problems, process their thoughts and feelings, and set goals [42, 48, 49]. In addition, other music therapy approaches and models, such as psychodynamic music therapy, insight-oriented music therapy, biomedical music therapy and vocal psychotherapy [42, 50], may be helpful when working with autoimmune neurological disorders.

These music therapy techniques are designed to improve a myriad of symptoms and needs by addressing the patients' intersectionalities [31, 38]. Table 10.1 provides examples and intervention ideas a music therapist may use for this population categorized by dimension and symptom or need.

Table 10.1 Intervention ideas for music therapy with patients with autoimmune neurological disorders by dimension

Physical dimension		
Symptom/need	Sample goal	Sample intervention idea
1. Difficulties with ambulation	Improve gait control/pace/movement	Rhythmic Auditory Stimulation (RAS®)
2. Speech distortion	Improve speech production	Singing and breathing exercises (e.g., Therapeutic Singing, Rhythmic Speech Cueing (RSC®), Oral Motor and Respiratory Exercises (OMREX®))
3. Chronic pain	Reduce pain perception	Guided imagery and breathing using music as a container for the experience
4. Deterioration of voice quality	Improve voice quality	Design vocal exercises through therapeutic singing

Cognitive dimension		
Symptom/need	Sample goal	Sample intervention idea
1. Memory issues	Maintain memory (short term)	Musical mnemonics training
2. Confusion	Improve executive function	Musical executive function training
3. Attention	Improve focus/attention	Musical attention and control training

Psychoemotional dimension		
Symptom/need	Sample goal	Sample intervention idea
1. Anxiety	Facilitate relaxation	Progressive muscle relaxation and music
2. Loss of autonomy	Increase perception of control	Call and response drumming. Patient chooses the instrument, sets the tempo, and develops the rhythm.
3. Coping with life changes	Develop coping mechanisms	Songwriting: Affirmation song. Encourage clients to use their song when they feel something in their lives is changing or they are facing a new/difficult situation.
4. Expressing thoughts and feelings	Increase verbalization of feelings and emotions	Invite the client to drum how they feel now.

Psychosocial dimension		
Symptom/need	Sample goal	Sample intervention idea
1. Isolation	Reduce perception of isolation	Social interaction through group music therapy sessions is recommended.
2. Changes in social roles/activities	Identify social strengths	Clients use musical instruments to play and represent social situations and their social strengths through instrumental performance.
3. Lack or changes in support network	Identify support network	Utilize a song that highlights the people that care about the patient in different aspects: family, friends, nurses, social workers, etc.

(continued)

Table 10.1 (continued)

Family/support system dimension			
	Symptom/need	Sample goal	Sample intervention idea
1.	Illness management	Increase understanding of the condition and treatment process	Psychoeducational MT session
2.	Changes of family roles	Redefine the patient's role within their family system	Songwriting: compare and contrast their role before and after illness, with emphasis on new perspectives these changes bring.
3.	Loss of ability to perform tasks as before	Identify personal strengths and abilities	Rewrite the lyrics of a patient selected song, where they can express their thoughts and feelings about their current abilities.

Spiritual dimension			
	Symptom/need	Sample goal	Sample intervention idea
1.	Finding meaning in/of life	Increase healthy outlook	Use lyric analysis of a song about gratitude to prompt patients to enlist things for which they are grateful and things to look forward to in life.
2.	Nurturing of inner self	Identify ways of meaningful self-care	Using a drum and a steady beat, the client chants a list of ways they can take care of themselves.
3.	Self-acceptance	Increase self esteem	Use songwriting and performance of a chant where the client completes "I am" statements (e.g., I am *important*).

Case Vignettes

Case 1

Alba is a woman in her 30s who was referred for private music therapy services by her neurologist after being diagnosed with demyelinating disease: multiple sclerosis. On the first music therapy session she appeared confused with all the information she had been given. She had a 1-month-old baby, and expressed feeling worried and helpless as she was actively breastfeeding her newborn. She was told to stop abruptly as the contrast given for her CT and MRI tests was excreted in her breast milk and that she would possibly need to stop breastfeeding altogether depending on the pharmacological treatment she was going to take. The patient cried as she was sharing her sadness every time she extracted her breast milk just to dispose of it. She also verbalized feeling very anxious and scared about the future.

The most imperative need was to provide a safe space and assist her in reducing her anxiety. The second immediate need identified was the need to learn coping skills that would help her process the wave of emotions she felt even after the session was over. The music therapist used a body-mind medicine approach [43] based on the biopsychosocial therapeutic model to induce relaxation. This consisted of guided breathing and relaxation supported by the music form and tempo used in the guitar. The music therapist's voice was soft but consistent, timed with the music to provide a sense of security, help the patient stay in the present, and used the ISO principle to determine the appropriate stimulus. A relaxation induction was used at the beginning of the intervention, followed by guided instructions to assist the patient in mindful body relaxation. As the intervention ended, the music therapist transitioned by using a gratitude chant. The client reported feeling calm and relaxed, and reported her perceived anxiety to be much lower. The last intervention of the session included identifying things Alba could do to feel supported during these times. She also identified at least three thoughts that may trigger her anxiety and thoughts to help orient herself to a calmer state. These positive thoughts were turned into affirmations such as "I am safe, I am love, I am at peace" and into a chant that the music therapist and the client sang, the music therapist played the guitar, and the client played the drums during this time. Alba's blood pressure and heart rate were lower after the music therapy session. She also appeared to be more relaxed and smiled often.

In the follow-up visits, these initial interventions were used as a foundation to scaffold future interventions designed to continue addressing the client's stress and anxiety while teaching her coping strategies to use as needed during the week.

Case 2

Lisa, a cisgender woman in her late 30s, was admitted to the hospital where a neurologist had issued an initial Huntington's disease diagnosis. She was then transferred to the short-term inpatient crisis management center from the hospital's neurological floor, where the psychiatrist was asked to confirm a secondary diagnosis of schizophrenia due to her aggression and unpredictable behavior. Lisa was later also diagnosed with autoimmune encephalitis (AE), which explained her neurological and psychiatric symptoms.

Lisa had difficulty processing information and communicating her thoughts and needs. She would have the urge to walk but lose the ability to walk halfway down the hallway. Sometimes she would stand up and not know how to sit back down. Lisa was confused, experienced hallucinations and delusions, and often talked to her partner through the room's television. For the most part, Lisa was calm and agreeable unless she felt threatened or felt that someone was interfering with her talking to her partner. Lisa's family parted ways after feeling overwhelmed by the complexity of the illness and what they perceived to be a change in her personality. Lisa had no support system, was unemployed, became unemployable, and uninsured. She was now in a psychiatric hospital without possibility of an appropriate neurological rehabilitation placement.

During daily music therapy sessions, the team initially provided diverse music to determine what would be familiar. We assessed by observing if Lisa would sing along or show any type of behavior to communicate liking or disliking the music. She became more engaged in group sessions once we discovered the music she enjoyed, often smiling, nodding, and sometimes singing along to some words. In approximately 2 months, Lisa communicated with thumbs up or down. Lisa became attentive to the strumming and began playing "air guitar". The team noticed how

Lisa observed and mimicked fast or slow strumming tempi patterns and happy and sad emotions while strumming the air guitar using facial expressions. In addition, she became tolerant of sitting through and listening to other patients' song choices. Lisa seemed less anxious and sometimes had more than one "good day" in a row. She enjoyed live music the most, even when not participating in playing the "air guitar". Interestingly, Lisa had no desire to hold a real guitar but was content with her "air guitar".

By the time Lisa was court-ordered and transferred to a psychiatric state hospital, her movements improved, and she demonstrated less aggressive behaviors. She communicated with yes or no responses, and the staff better understood her needs. Although the receiving state hospital offered music therapy, as a newcomer, Lisa was placed on a waiting list.

Conclusion and Future Considerations

While there is limited research on music therapy for autoimmune neurological disorders, there is relevant published information for music therapists to make informed decisions when assessing and treating patients with these conditions. Music therapy interventions can be designed to assist the patient from the moment symptoms appear, through the process of diagnosis, and throughout treatment. Music therapy research provides opportunities to use effective evidence-informed interventions treating many conditions, including symptoms such as those in the autoimmune neurological disorders and diseases area [31].

A music therapist working with patients with autoimmune neurological disorders must consider the patient's own expression of the diagnosis in order to conduct an accurate assessment of needs and create an effective treatment plan [31]. Pacing is very important when working with clients needing extra time to move, talk, or process - so silence in music should be part of the process. An effective use of music may provide opportunities for the client to have increased perception feelings of control and autonomy. Providing clients with options on their level of participation and ability, and allowing for autonomy through decision-making, such as selecting songs and instruments, will reiterate that their needs and wants are a priority [51, 52].

There is a great need to expand the scope of music therapy research related to autoimmune neurological diseases to ensure that those patients access the appropriate music therapy treatment and related resources. In parallel with research progress in autoimmune neurological disorders, music therapy and music-based intervention research needs to develop its evidence base. Music therapists should continue educating other professionals on how music can impact the many body systems affected by these conditions.

References

1. Autoimmune diseases [Internet]. National Institute of Environmental Health Sciences. U.S. Department of Health and Human Services. 2023 [cited 2023 Feb 25]. Available from: https://www.niehs.nih.gov/health/topics/conditions/autoimmune/index.cfm
2. Bhagavati S. Autoimmune disorders of the nervous system: pathophysiology, clinical features, and therapy. Front Neurol. 2021;12 https://doi.org/10.3389/fneur.2021.664664.
3. McKeon A. Autoimmune brain diseases: what you need to know. Lancet Neurol. 2022;21(10):875. https://doi.org/10.1016/s1474-4422(22)00354-4.
4. Dalmau J, Graus F. Autoimmune encephalitis and related disorders of the nervous system. 2022; https://doi.org/10.1017/9781108696722.
5. Harrison K. New report reveals that, while undiagnosed, rare disease patients have cost the NHS in excess of £3.4 billion [Internet]. Imperial College Health Partners. 2018 [cited 2023 Apr 4]. Available from: https://imperialcollegehealthpartners.com/new-report-reveals-undiagnosed-rare-disease-patients-cost-nhs-excess-3-4-billion/
6. Lerner A, Jeremias P, Matthias T. The world incidence and prevalence of autoimmune diseases is increasing. Int J Celiac Dis. 2016;3(4):151–5. https://doi.org/10.12691/ijcd-3-4-8.
7. Zivadinov R, Raj B, Ramanathan M, Teter B, Durfee J, Dwyer MG, et al. Autoimmune comorbidities are associated with brain injury in multiple sclerosis. Am J Neuroradiol. 2016;37(6):1010–6. https://doi.org/10.3174/ajnr.A4681.
8. Rubin DB, Batra A, Vaitkevicius H, Vodopivec I. Autoimmune neurologic disorders. Am J Med. 2018;131(3):226–36. https://doi.org/10.1016/j.amjmed.2017.10.033.
9. Chandrashekara S. The treatment strategies of autoimmune disease may need a different approach from conventional protocol: a review. Indian J Pharmacol. 2012;44(6):665. https://doi.org/10.4103/0253-7613.103235.
10. Waubant E. Improving outcomes in multiple sclerosis through early diagnosis and effective management. Prim Care Companion CNS Disord. 2012; https://doi.org/10.4088/pcc.11016co2cc.
11. Epstein NE. Multidisciplinary in-hospital teams improve patient outcomes: a review. Surg Neurol Int. 2014;5(8):295–303. https://doi.org/10.4103/2152-7806.139612.
12. Cullen AE, Holmes S, Pollak TA, Blackman G, Joyce DW, Kempton MJ, et al. Associations between non-neurological autoimmune disorders and psychosis: a meta-analysis. Biol Psychiatry. 2019;85(1):35–48. https://doi.org/10.1016/j.biopsych.2018.06.016.
13. Kayser MS, Dalmau J. The emerging link between autoimmune disorders and neuropsychiatric disease. J Neuropsychiatr. 2011;23(1):90–7. https://doi.org/10.1176/appi.neuropsych.23.1.90.
14. Pape S, Gevers TJG, Vrolijk JM, van Hoek B, Bouma G, van Nieuwkerk CMJ, et al. Rapid response to treatment of autoimmune hepatitis associated with remission at 6 and 12 months. Clin Gastroenterol Hepatol. 2020;18(7) https://doi.org/10.1016/j.cgh.2019.11.013.
15. Flanagan EP, Geschwind MD, Lopez-Chiriboga AS, Blackburn KM, Turaga S, Binks S, et al. Autoimmune encephalitis misdiagnosis in adults. JAMA Neurol. 2023;80(1):30. https://doi.org/10.1001/jamaneurol.2022.4251.
16. Abbatemarco JR, Rodenbeck SJ, Day GS, Titulaer MJ, Yeshokumar AK, Clardy SL. Autoimmune neurology. Neurol Neuroimmunol Neuroinflamm. 2021;8(5)
17. Boeschoten RE, Braamse AMJ, Beekman ATF, Cuijpers P, van Oppen P, Dekker J, et al. Prevalence of depression and anxiety in multiple sclerosis: a systematic review and meta-analysis. J Neurol Sci. 2017;372:331–41. https://doi.org/10.1016/j.jns.2016.11.067.
18. American Psychiatric Association. Diagnostic and statistical manual of mental disorders. 5th ed TR(tm); 2022.
19. Johnston ME, Wallace LE. Stress and medical procedures. Oxford University Press; 1990.
20. Medical trauma [Internet]. 2023 [cited 2023 Apr 1]. Available from: https://istss.org/public-resources/friday-fast-facts/fast-facts-medical-trauma#:~:text=Medical%20trauma%20is%20emotional%20and,be%20difficult%2C%20uncomfortable%20or%20frightening

21. McEwen BS. Neurobiological and systemic effects of chronic stress. Chronic Stress. 2017;1 https://doi.org/10.1177/2470547017692328.
22. Davis WB. Introduction to music therapy: theory and practice. 3rd ed. McGraw-Hill Humanities, Social Sciences & World Languages; 2008.
23. Korhan EA, Uyar M, Eyigör C, Hakverdioğlu Yönt G, Çelik S, Khorshıd L. The effects of music therapy on pain in patients with neuropathic pain. Pain Manag Nurs. 2014;15(1):306–14. https://doi.org/10.1016/j.pmn.2012.10.006.
24. Vicznesky AM. Is music therapy effective at reducing pain in fibromyalgia patients? [Internet]. DigitalCommons@PCOM. 2019 [cited 2023 Feb 25]. Available from: https://digitalcommons.pcom.edu/pa_systematic_reviews/437/
25. Hasanah I, Mulatsih S, Haryanti F, Haikal Z. Effect of music therapy on cortisol as a stress biomarker in children undergoing IV-line insertion. J Taibah Univ Med Sci. 2020;15(3):238–43. https://doi.org/10.1016/j.jtumed.2020.03.007.
26. Ventegodt S, Kandel I, Ervin DA, Merrick J. Concepts of holistic care. In: Health care for people with intellectual and developmental disabilities across the lifespan. Berlin: Springer; 2016. p. 1935–41.
27. American Psychology Association. Coping with a diagnosis of chronic illness [Internet]. American Psychological Association. 2013 [cited 2023 Apr 5]. Available from: https://www.apa.org/topics/chronic-illness/coping-diagnosis
28. Lopez N, Gadsden VL. Health inequities, social determinants, and intersectionality. NAM Perspect. 2016;6(12) https://doi.org/10.31478/201612a.
29. Chovatiya R, Medzhitov R. Stress, inflammation, and defense of homeostasis. Mol Cell. 2014;54(2):281–8. https://doi.org/10.1016/j.molcel.2014.03.030.
30. Katz JD. At the intersection of self and not-self: Finding the locus of 'self' in autoimmunity [Internet]. Medical humanities. U.S. National Library of Medicine. 2018 [cited 2023 Apr 1]. Available from: https://www.ncbi.nlm.nih.gov/pmc/articles/PMC6087542/
31. Hanser SB. The new music therapist's handbook. 3rd ed. Boston: Berklee Press; 2019.
32. Boiko EA, Ivanchuk EV, Gunchenko MM, Batysheva TT. The potential of music therapy in neurology using multiple sclerosis as an example. Neurosci Behav Physiol. 2017;47(5):570–2. https://doi.org/10.1007/s11055-017-0437-8.
33. Lopes J, Keppers II. Music-based therapy in rehabilitation of people with multiple sclerosis: a systematic review of clinical trials. Arq Neuropsiquiatr. 2021;79(6):527–35. https://doi.org/10.1590/0004-282x-anp-2020-0374.
34. Magee WL, Davidson JW. Music therapy in multiple sclerosis: results of a systematic qualitative analysis. Music Ther Perspect. 2004;22(1):39–51. https://doi.org/10.1093/mtp/22.1.39.
35. Ostermann T, Schmid W. Music therapy in the treatment of multiple sclerosis: a comprehensive literature review. Expert Rev Neurother. 2006;6(4):469–77. https://doi.org/10.1586/14737175.6.4.469.
36. Vinciguerra C, De Stefano N, Federico A. Exploring the role of music therapy in multiple sclerosis: brief updates from research to clinical practice. Neurol Sci. 2019;40(11):2277–85. https://doi.org/10.1007/s10072-019-04007-x.
37. Schmid W, Aldridge D. Active music therapy in the treatment of multiple sclerosis patients: a matched control study. J Music Ther. 2004;41(3):225–40. https://doi.org/10.1093/jmt/41.3.225.
38. Thaut MH, McIntosh GC, Hoemberg V. Neurologic music therapy: from social science to neuroscience. In: Handbook of neurologic music therapy. Oxford University Press; 2014. p. 1–6.
39. Pakdeesatitwara N, Tamplin J. Music therapy services in neurorehabilitation: an international survey. Aust J Music Ther. 2018;29:63–90.
40. Raglio A. Effects of music and music therapy on mood in neurological patients. World J Psychiatry. 2015;5(1):68. https://doi.org/10.5498/wjp.v5.i1.68.
41. Impellizzeri F, Leonardi S, Latella D, Maggio MG, Foti Cuzzola M, Russo M, et al. An integrative cognitive rehabilitation using neurologic music therapy in multiple sclerosis. Medicine. 2020;99(4) https://doi.org/10.1097/md.0000000000018866.
42. Darrow A-A. Introduction to approaches in music therapy. American Music Therapy Association: Silver Spring; 2008.

43. The Center for Body-Mind Medicine. The Center for Body-Mind Medicine [Internet]. CMBM. 2023 [cited 2023 Apr 6]. Available from: https://cmbm.org/
44. Fancourt D, Ockelford A, Belai A. The psychoneuroimmunological effects of music: a systematic review and a new model. Brain Behav Immun. 2014;36:15–26.
45. Taylor DB. Biomedical foundations of music as therapy. Eau Claire: Barton Publications; 2018.
46. González-Díaz SN, Arias-Cruz A, Elizondo-Villarreal B, Monge-Ortega OP. Psychoneuroimmunoendocrinology: clinical implications. World Allergy Organ J. 2017;10:19. https://doi.org/10.1186/s40413-017-0151-6.
47. Chagani SM. Psychoeducation – a mental health promoting practice in schizophrenia. J Pioneer Med Sci. 2015;5:3–6.
48. Silverman MJ. Music therapy in mental health for illness management and recovery. Oxford: Oxford University Press; 2022.
49. Thaut MH, Hoemberg V. Handbook of neurologic music therapy. Oxford: Oxford University Press; 2016.
50. Thaut MH. Music as therapy in early history. Prog Brain Res. 2015:143–58.
51. Borczon RM. Music therapy: a fieldwork primer. Dallas: Barcelona Publishers; 2017.
52. Entwistle VA, Carter SM, Cribb A, McCaffery K. Supporting patient autonomy: the importance of clinician-patient relationships. J Gen Intern Med. 2010;25(7):741–5. https://doi.org/10.1007/s11606-010-1292-2.

Music for Epilepsy

Robert J. Quon, Ondřej Strýček, Alan B. Ettinger,
Michael A. Casey, Ivan Rektor, and Barbara C. Jobst

Introduction

This chapter will discuss the current knowledge about the use of music-based interventions (music therapy and music medicine) for adults and children with epilepsy. We review the epidemiology of epilepsy, discuss possible mechanisms for the effects of music on clinical seizures and electroencephalographic activity, discuss the specific effects of Mozart's music on seizures (including theories regarding the "Mozart effect"), and outline key future research directions to improve our understanding of how music can help individuals with seizures.

R. J. Quon (✉)
Warren Alpert Medical School of Brown University, Providence, RI, USA
e-mail: robert_quon@brown.edu

O. Strýček · I. Rektor
First Department of Neurology, St. Anne's University Hospital and Faculty of Medicine, Masaryk University, Brno, Czech Republic

Multimodal and Functional Neuroimaging Research Group, CEITEC-Central European Institute of Technology, Masaryk University, Brno, Czech Republic
e-mail: ondrej.strycek@fnusa.cz; irektor@med.muni.cz

A. B. Ettinger
United Diagnostics and United Medical Monitoring, New Hyde Park, NY, USA

Professional Advisory Board of the EPIC/Epilepsy Foundation of Long Island,
East Meadow, NY, USA

M. A. Casey
Department of Music and Computer Science, Dartmouth College, Hanover, NH, USA
e-mail: michael.a.casey@dartmouth.edu

B. C. Jobst
Department of Neurology and Neurocritical Care, Dartmouth Health, Lebanon, NH, USA

Geisel School of Medicine at Dartmouth, Hanover, NH, USA
e-mail: barbara.c.jobst@hitchcock.org

© The Author(s), under exclusive license to Springer Nature Switzerland AG 2023
K. Devlin et al. (eds.), *Music Therapy and Music-Based Interventions in Neurology*, Current Clinical Neurology,
https://doi.org/10.1007/978-3-031-47092-9_11

Self-Location

Jobst Group Perspectives

The Jobst group focuses on understanding how music influences the brains of patients with medication-refractory epilepsy. Their goal is to identify musical components essential for creating beneficial responses in epilepsy, then to utilize those components to engineer auditory stimuli that may serve as adjunctive therapies.

Rektor Group Perspectives

The Rektor group focuses on research of therapeutic aspects of music in medication-refractory epilepsy using intracranial electroencephalography (iEEG). Their research specializes in analyzing acoustic properties of music and their possible therapeutic effect. Their aim is to create individualized, tailored music interventions for patients with epilepsy.

Definitions and Background

Epilepsy is one of the most common chronic neurologic conditions with over 3.4 million US residents living with some form of epilepsy and over 150,000 persons newly diagnosed with epilepsy annually [1]. Epilepsy is characterized as a disease of the brain leading to repeated, unprovoked seizures [2]. It is a heterogeneous condition encompassing multiple seizure types and syndromes that arise from numerous etiologies [3]. Antiseizure medication (ASM) therapy is generally started after two or more unprovoked seizures (or after one seizure if there is a high probability of seizure recurrence), resulting in approximately 60–70% of persons with epilepsy achieving seizure freedom [4]. Despite the development of many new ASMs over recent years, this – unfortunately – translates to as many as 30% failing to achieve seizure freedom after two or more ASM trials; these persons are classified as having medication-resistant or refractory epilepsy [5].

The only curative treatment option for persons with refractory epilepsy is surgical resection. Surgical candidates are people with medication-resistant focal epilepsy that have seizures originating in brain regions that can be removed with minimal risk of secondary dysfunction. This excludes persons with bilateral or multifocal seizure onset, generalized onset epilepsy, comorbidities that prohibit surgery, and seizure propagation closely linked with eloquent cortex [6]. For those not eligible for surgery, other palliative treatments are aimed at reducing the seizure burden and improving quality of life. These include neuromodulation (vagus nerve stimulation [VNS], deep brain stimulation [DBS], responsive cortical stimulation) [7, 8], additional trials of other ASMs, and lifestyle modifications (e.g., ketogenic diet) [9]. Thus, developing novel, well-tolerated supplementary therapies for refractory epilepsy remains a key area of need.

Interictal Epileptiform Discharges

Ictal activity (seizure activity) is the ideal outcome for research studies; however, seizures are difficult to predict. It would also require long-term exposure to interventions while offering limited control of other experimental variables. Fortunately, there are several established electrophysiological biomarkers of epileptogenesis and epilepsy. These make it easier to identify and validate new interventions that can potentially reduce epileptic seizures. One of the most widely accepted biomarkers of epilepsy is interictal epileptiform discharges (IEDs), which consist of spikes (20–70 ms duration) and sharp waves (70–200 ms duration). IEDs represent the transient hypersynchronous firing of hyperexcitable populations of neurons and are specific to persons with epilepsy [10].

There are two competing hypotheses regarding the relationship between IEDs and ictogenesis. The first is that IEDs are beneficial for persons with epilepsy, as past studies demonstrated a decrease in IED rates prior to a seizure [11]. It is speculated that the protective function of IEDs lies in the hyper-polarizations that follow IEDs that may inhibit aberrant activity [12]. IEDs are also thought to generate a chronic low level of excitability that restricts the generation of seizures [13]. A more popular hypothesis, with increasing support over the years, is that IEDs are detrimental to persons with epilepsy. This is based on findings that IEDs result from cyclical increases in neural excitability, which result in a seizure once a critical density is achieved [14]. In addition, IEDs were shown to have a causal relationship with seizure generation [13]. Irrespective of why IEDs are generated, there is support for the association between IEDs and other adverse health outcomes, especially cognitive impairment, and an impetus to reducing IEDs in persons with epilepsy [15] (Fig. 11.1).

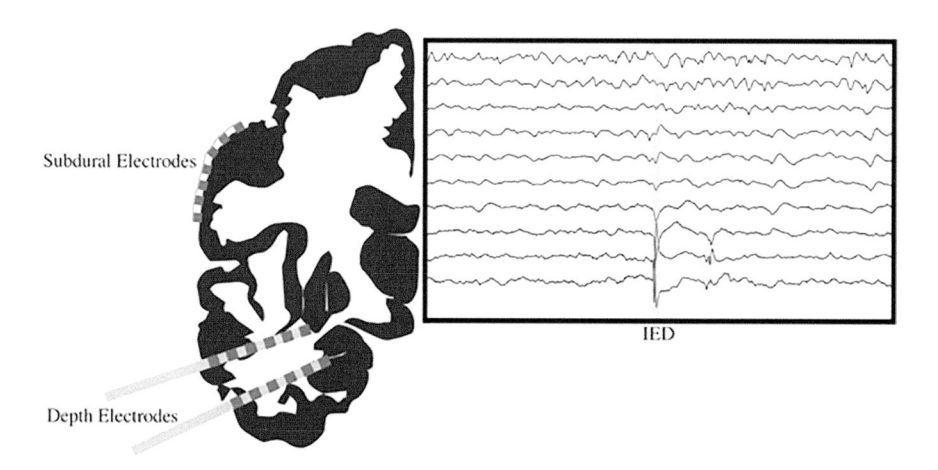

Fig. 11.1 Invasive monitoring of epileptiform activity. Different intracranial electrodes and a sample interictal epileptiform discharge (IED) detected with intracranial electroencephalography (EEG)

Literature Review

Of the nonpharmacologic interventions for refractory epilepsy, there is growing interest in the utilization of music as an adjunctive therapy. A general limitation in the field is the heterogeneity of study designs and small population samples. Pertinent studies discussed below are stratified based on the outcome of interest (i.e., interictal or ictal activity). We will briefly discuss popular theories regarding why music may benefit persons with epilepsy, then the following sections will discuss evidence for the benefits of music for epilepsy.

Theories About Why Music May Help Epilepsy

The action observation/execution matching system or the "human mirror neuron system" may contribute to the beneficial effects of music by linking exogenous auditory stimulation with the motor cortex, which is connected to pathways that may modulate other neural activity [16]. It is proposed that the act of experiencing music is related to the perception of hierarchically organized motor elements with temporally synchronous auditory information [17]. The anterior insula may link this mirror neuron system with the limbic system to generate a complex affective response to music [17]. It is theorized that this linked response allows the sender to generate a "shared" musical experience with the perceiver, evoking emotional and autonomic reactions. It was also proposed that enhanced parasympathetic tone secondary to music exposure is linked with decreased epileptiform activity. This is a promising mechanism to explore given prior evidence of reduced seizure activity with parasympathetic nervous system stimulation (vagus nerve stimulation, FDA-approved for treatment-refractory seizures) [18].

The effects of music could also be explained by its influence on the release of neurotransmitters, such as GABA and dopamine [19]. GABA interneurons serve as regulators for excitatory and inhibitory neural activity, synchronizing cortical pyramidal neurons to fire at gamma frequencies (range 30–100 Hz; 40 Hz emphasized in research) [20]. Provided that epileptic networks are associated with GABA dysfunctions [21], and findings that exogenous auditory stimuli may entrain patterns of cortical gamma [22], it follows that this may be a potential path for beneficial effects. Alternatively, a Positron emission tomography (PET) study showed increased dopamine release was associated with emotional responses to music [23], while dopamine dysregulation was linked with epilepsy [24]. Thus, music may help reverse dopamine abnormalities concurrent with epilepsy. It is also hypothesized that alterations of dopamine receptors/dopamine binding capacity in the basal ganglia may play a role in several types of epilepsy [25], while basal ganglia connectivity with the cortex plays a vital inhibitory role in focal epilepsy [26]. This is intriguing, as listening to music is believed to modify dopaminergic pathways within subcortical structures to influence thalamocortical projections [27].

The best-known theory regarding the potential anticonvulsive effects of music is the "Mozart effect", which will be discussed in detail in this chapter. The "Mozart effect" was coined by Rauscher et al. [28] when they observed a short-term enhancement of spatial-temporal reasoning in college students with exposure to 10 min of Mozart's Sonata for Two Pianos in D Major (Mozart's K448) [28]. Subsequent animal studies showed that music exposure might enhance dendritic branching and neurogenesis in the amygdala and hippocampus [29]. Despite growing scientific skepticism regarding the existence of a lasting and reproducible "Mozart effect", these promising observations led to an exploration of the "Mozart effect" for various neurological conditions.

It is hypothesized that the "complex and highly organized architecture" of Mozart's K448 may resonate with the super organization of the cerebral cortex, as demonstrated by the Trion model [30]. The Trion model is a mathematical theory of information-processing and memory recall based on Mountcastle's columnar organization principle of the cortex [31], whereby exposure to a highly organized auditory stimulus across time and space results in the reorganization of cortical firing; it is theorized that these learned patterns may suppress epileptogenic activity [32]. In addition to oscillatory patterns and organization, timbre (tone color or perceived sound quality) and sound energy (movement of energy through a substance in waves) of Mozart's music were shown to contribute to beneficial effects [33].

Several prior studies demonstrated that persons with epilepsy have reported a subjective indifference to Mozart's K448 and classical music in general [34, 35]. Rafiee et al. [36] proposed a cortical dynamics theory [36], which posited that if the 1/frequency (1/f) scaling of music and the brain "match", the two resonate together, resulting in the maximal exchange of information. This builds on the findings of Borges and colleagues [37], suggesting that classical music influences neural activity by a 1/f resonance mechanism [37]. However, this 1/f theory is distinct from the theory of musical entrainment, which suggests that cortical oscillations become phase-locked to note onsets within the musical stimuli [38, 39]. A theory linking several of the past hypotheses is focused on positive reward prediction errors. Arjmand et al. [40] lend support to this theory by showing that unexpected changes in musical features (e.g., intensity, tempo) preferentially activated frontal cortical brain regions associated with positive emotional responses [40]. Thus, violations in frontal cortex predictions may activate emotion processing networks, leading to the suppression of aberrant cortical activity [35, 41]. Overall, these theories motivate the continued exploration of the therapeutic benefits of music for epilepsy.

Music and Interictal Epileptiform Discharges

Hughes et al. [42] were the first to show a reduction in interictal epileptiform discharges (IEDs) while persons with epilepsy listened to Mozart's K448 [42]. A prospective, randomized, single-masked crossover, placebo-controlled, pilot clinical trial by Turner [43] later demonstrated a significant reduction in IEDs during Mozart's K448 in two of four subjects with epilepsy [43]. The IED-reducing effect of Mozart's K448 was later confirmed by other scalp-EEG studies of pediatric and adult populations in randomized controlled trials (RCTs) [44, 45], prospective studies [16, 18, 32, 33, 46, 47], and meta-analyses [48, 49]. Moreover, the same beneficial effect was observed with intracranial EEG (iEEG), a modality that provides optimal neural signals with the highest sensitivity for detecting true interictal activity [34, 35].

Mozart's Piano Sonata in C Major ("Mozart's K545") was the only other composition to show a significant reduction in IEDs, though the strongest evidence to date belongs to Mozart's K448 [46, 50]. Few other protocols directly compared Mozart's K448 to the music of other composers or other genres. Music tested in the past includes Beethoven's Fur Elise, a digitally created string version of K448, Wagner's Lohengrin, age-related control music for children (e.g., Busted's Year 3000), and Haydn's Symphony No. 94 (Surprise Symphony) [32, 34, 35, 43]. Quon et al. [35] also examined Chopin's Bolero in C – Op. 19, Liszt's Piano Sonata in B Minor (first movement: Lento assai – *Allegro energico*), and various songs from a patient-preferred musical genre (e.g., Judas Priest's Jugulator and Buddy Holly's Peggy Sue) [35]. Surprisingly, none of these other musical stimuli demonstrated significant reductions in IEDs – apart from Haydn's symphony, which reduced IEDs in women and increased IEDs in men [34]. This suggests that the stimuli may have to be tailored to characteristics such as gender and generation. At the same time, it indicates that there might be something specific to the music parameters (e.g., organization, predictability) inherent to music of the classical period that is at play. It remains unknown whether there is a differential effect of vocal versus instrumental music for epilepsy.

Why Mozart's sonatas (K448 and K545) are unique in reducing IEDs remains a matter for discussion. A promising explanation is that Mozart's K448 contains unique long-term periodicities that differ from other compositions due to its higher repetition rate. It is theorized that these long-term periodicities, lasting 10–60 s, likely resonate with brain activity [30, 35]. Rafiee et al. [36] revealed the importance of these long-term periodicities in showing no therapeutic effect after destroying the temporal components of Mozart's K448 via shuffling the phase of each frequency randomly but preserving the power spectrum of the original piece [36]. The IED-reducing effect was also absent with a digitally computerized version of Mozart's K448 that was rendered for a string quartet. This suggests that while the repetition or periodicity of harmonies contributes to the IED-reducing effect, various other characteristics are likely involved. For instance, Haydn's Symphony No. 94 contains a repetition of harmonies like Mozart's K448, especially relative to music from other classical composers (e.g., Bach, Wagner, Beethoven, Chopin, and Liszt); however, it did not have the same IED-reducing effect [30, 34, 35]. These preliminary findings encourage future research focused on identifying additional characteristics to create patient-centered and personalized music interventions.

Apart from repetition, researchers are studying other musical components that may be important for the beneficial effect music has on IEDs. Štillová et al. [34] showed that non-dissonant music with a harmonic spectrum, decreasing tempo, and significant high-frequency parts was associated with IED reductions in men while being gradually less dynamic in terms of loudness had a major impact on IEDs in women [34]. They showed that both men and women reacted positively to tempo and tonal quality, independent of patient age, yet the precise mechanism by which men and women respond differently remains unknown [34]. Cognitive function was shown to predict treatment response, as music was more effective in treating children with an IQ >70; however, this correlation with IQ was not significant in other studies [16]. Recently, Quon et al. [35] utilized music information retrieval (MIR) techniques, adapted from McFee and Ellis [51], to quantitatively assess patterns within Mozart's K448 [35, 51]. In doing so, they confirmed that a complex amalgamation of features likely contributes to the IED-reducing effects seen with Mozart's K448. Using artificial intelligence to detect or enhance these features in other pieces of music holds the potential for identifying other IED-reducing pieces.

From a therapeutic perspective, one of the cardinal questions is how long and frequent music exposure should last to effectively reduce IEDs or seizures. In some studies, the reduction in IEDs was evident after only 30 s of exposure [35] and after a single session of music stimulation [49]. The persistence of this beneficial effect (i.e., carry-over effect) was shown to depend on the duration of music. It was not seen after 90 s of exposure but was evident after listening to Mozart's K448 for several minutes [34, 35, 42]. The meta-analysis by Sesso and Sicca [49] showed a 28% reduction in IEDs when listening to Mozart's K448 compared to baseline and a 20% reduction in IEDs after a single stimulation session.

Music and Seizures

Despite promising results supporting the use of music for IEDs, evidence for music's therapeutic effect on seizures remains weak. As mentioned above, IEDs are recognized as biomarkers for epilepsy. However, it is of great clinical interest if music may reduce clinical seizures. A few case reports show music's direct effect on seizure recurrence [52–54]. Lahiri and Duncan [52] report a case of a 56-year-old man with refractory gelastic epilepsy (laughing fits) and secondarily generalized tonic-clonic seizures that occurred seven times per month, on average. After listening to Mozart for 45 min per day for 3 months, the patient ceased to have any secondarily generalized tonic-clonic seizures; he continued to have gelastic seizures, but the manifestations were much milder than before listening to Mozart (5–9 s of brief smiling) [52].

Another group reported two cases of patients with refractory nonconvulsive status epilepticus and a positive response to music. The first was a 22-year-old man with persistent impairment of consciousness secondary to nonconvulsive status epilepticus that developed after severe brain trauma (Glasgow Coma Scale score of 4) [53]. The patient failed multiple antiseizure medication trials, then showed improvement in electrographic epileptic discharges after 5 days of Mozart's K448 for 30 min daily. The patient emerged from a coma and experienced continued cognitive and motor functioning improvements while listening to Mozart's K448 for 1 h daily [53]. The second case is of a 68-year-old man with refractory nonconvulsive status epilepticus secondary to brain metastases from lung cancer [54]. The patient was in this state for 1 week and failed multiple antiseizure medication trials. The electrographic status epilepticus remitted within 8 h of continuous exposure to the patient's preferred music (different works by Mozart and J.S. Bach), with no medication changes during this period [54]. These reports suggest music may be a potential adjunctive therapy for patients with refractory status epilepticus.

Nevertheless, there is limited research investigating patient-preferred music relative to specific Mozart compositions with seizure reduction as the outcome. Rafiee et al. [55] suggested that there must be something special about the original structure of Mozart's K448 for seizure reduction [55]. In a randomized crossover study (n = 13, age range = 26–75), they exposed patients to the first 6 min of Mozart's K448 for 3 months and a phase-scrambled version of Mozart's K448 for 3 months. They found a reduction in seizures with the original version of Mozart's K448 but not the phase-scrambled version [55]. Additionally, Coppola et al. [46] showed that 45.4% of patients with drug-resistant epileptic encephalopathy and severe intellectual disability secondary to cerebral palsy (n = 11, age range = 1.5–21 years) had a > 50% reduction in the total number of seizures after listening to Mozart's compositions (Symphony no. 41, k551; Piano Concerto no. 22, k482; Violin Concerto no. 1, k207; Violin Concerto no. 4 in D major, K218, *Allegro aperto*; Symphony no. 46 in C major, kv96, *Allegro*; Flute Concerto in D major K314, *Allegro aperto*) for 2 h per day for 15 days [46]. All subjects had a decrease in the total number of seizures compared with baseline (51.5% reduction during the 15-day music intervention; 20.7% reduction 2 weeks following the music intervention) [46]. In both studies, subjects could decide on the timing of the music intervention each day.

Several randomized controlled trials also showed the seizure-reducing potential for music [44, 45, 47, 55, 56]. Bodner et al. [44] conducted a three-year RCT in patients with epilepsy or seizures related to an underlying condition (n = 73, age range = 12–78 years) [44]. They showed that subjects exposed to Mozart's K448 for 10 h nightly (9:00 pm to 7:00 am) for a year had a significant 24% seizure-reduction during the treatment phase and 33% seizure reduction during the 1-year follow-up phase; 24% of subjects were seizure-free during the 1-year treatment period [44]. A similar effect was shown by D'Alessandro et al. [56], who used a randomized crossover design (n = 12, mean age 21.6 years), exposing patients to Mozart's K448 once daily for 6 months (timing of exposure unspecified) [56]. The average seizure reduction was 20.5%; one patient was seizure-free, and five had a greater than 50% reduction in seizure frequency. Mozart's K448 even reduced seizures with exposure once daily before bedtime for at least 6 months [45]. The seizure medication schedule was unaltered during the data collection period of these studies.

Interestingly, Coppola et al. [47] demonstrated that there was an even greater seizure reduction while listening to a set of Mozart's music (77% seizure reduction) compared to Mozart's K448 alone (37% seizure reduction) [47]. They had pediatric patients with drug-resistant epileptic encephalopathies or syndromes (n = 19, age range 1–24 years) listen to either Mozart's K448 or a set of Mozart's compositions for 2 h per day for 2 weeks [47]. Listening time was distributed over the day based on subject preferences. Together, these studies demonstrate consistent results with music interventions and suggest that music may be utilized as a potential neuromodulation technique for seizures.

Music and Epilepsy Comorbidities

In addition to seizures, persons with epilepsy have comorbidities that drastically impact their quality of life [57]. For instance, persons with epilepsy experience much higher levels of stress due to the unpredictable nature of seizure occurrences [58]. Stress may, in turn, lower the seizure threshold, precipitating a vicious cycle [59]. Stress was shown to impair memory and cognition, which remain among the top complaints in persons with epilepsy [60]. There is also a higher frequency of mood disorders comorbid with epilepsy. These mood disorders are related to various etiologies and may be exacerbated by the general loss of independence (e.g., inability to drive, higher incidence of unemployment) that comes with the diagnosis [61]. Quality of life remains reduced even after seizure control, as antiseizure medications may have an adverse effect on mood, potentially leading to medication-induced depression or psychosis [62]. Thus, future research on the use of music medicine and music therapy to target these comorbid conditions may drastically improve the quality of life of persons with epilepsy.

It is important to differentiate music therapy from music-based interventions discussed in this chapter. Music therapy involves a credentialed professional trained in specific techniques to target psychological, behavioral, and cognitive symptoms – in addition to interpersonal relationships. They strive to improve the psychosocial well-being of those coping with epilepsy [63]. One study showed that music therapy administered during video EEG testing – something done for the routine workup of epilepsy – helped the coping ability of paediatric patients [64]. Despite its use in other contexts, music therapy has not been widely applied to epilepsy. Further research is needed to confirm the benefits of music therapy for addressing psycho-behavioral, cognitive, and other aspects of epilepsy.

Conclusion and Future Considerations

There is growing support for the use of music as an intervention to reduce ictal and interictal epileptiform activity in persons with epilepsy. A general limitation in the field is the use of small, heterogeneous samples and varying protocols. Most past studies also rely on scalp EEG, which is sub-optimal for detecting the full extent of neural activity. Despite these limitations, there is growing evidence for the benefits of Mozart K448 in epilepsy, especially regarding its IED-reducing capabilities. Future research is encouraged to determine how and why Mozart's K448 is beneficial for persons with epilepsy and whether there are other compositions showing similar effects. The question of whether specific music such as Mozart's K448 should be broadly utilized across different individuals suffering from epilepsy or whether patient preference should be the primary consideration in music selection remains unanswered. Large multicenter studies using homogeneous samples and protocols are needed to help address these questions. The evidence presented in this chapter encourages the continued exploration of music-based neuromodulation as a non-pharmacologic, adjunctive intervention for epilepsy.

References

1. Zack MM, Kobau R. National and state estimates of the numbers of adults and children with active epilepsy – United States, 2015. MMWR Morb Mortal Wkly Rep. 2017;66(31):821–5.
2. Beghi E. The epidemiology of epilepsy. Neuroepidemiology. 2020;54(Suppl. 2):185–91.
3. Duncan JS, Sander JW, Sisodiya SM, Walker MC. Adult epilepsy. Lancet. 2006;367:1087–100.
4. Bonnett LJ, Tudur Smith C, Donegan S, Marson AG. Treatment outcome after failure of a first antiepileptic drug. Neurology. 2014;83(6):552–60.
5. Kwan P, Schachter S, Brodie M. Drug-resistant epilepsy. N Engl J Med. 2011;365:919–26.
6. Khoo A, Tisi J, Mannan S, O'Keeffe AG, Sander JW, Duncan JS. Reasons for not having epilepsy surgery. Epilepsia. 2021;62(12):2909–19.
7. Morrell MJ. Responsive cortical stimulation for the treatment of medically intractable partial epilepsy. Neurology. 2011;77(13):1295–304.
8. Fisher R, Salanova V, Witt T, Worth R, Henry T, Gross R, et al. Electrical stimulation of the anterior nucleus of thalamus for treatment of refractory epilepsy: deep brain stimulation of anterior thalamus for epilepsy. Epilepsia. 2010 May;51(5):899–908.
9. Prasad AN, Stafstrom CF, Holmes GL. Alternative epilepsy therapies: the ketogenic diet, immunoglobulins, and steroids. Epilepsia. 1996;37(s1):S81–95.
10. Jabbari B, Russo MB, Russo ML. Electroencephalogram of asymptomatic adult subjects. Clin Neurophysiol. 2000;4
11. Gotman J. Relationships between interictal spiking and seizures: human and experimental evidence. Can J Neurol Sci J Can Sci Neurol. 1991 Nov;18(S4):573–6.
12. Matsumoto H, Marsan CA. Cortical cellular phenomena in experimental epilepsy: ictal manifestations. Exp Neurol. 1964;9:305–26.
13. Avoli M, Biagini G, de Curtis M. Do interictal spikes sustain seizures and epileptogenesis? Epilepsy Curr. 2006 Nov;6(6):203–7.
14. Karoly PJ, Freestone DR, Boston R, Grayden DB, Himes D, Leyde K, et al. Interictal spikes and epileptic seizures: their relationship and underlying rhythmicity. Brain. 2016;139(4):1066–78.
15. Kleen JK, Scott RC, Holmes GL, Roberts DW, Rundle MM, Testorf M, et al. Hippocampal interictal epileptiform activity disrupts cognition in humans. Neurology. 2013;81(1):18–24.
16. Lin LC, Lee WT, Wu HC, Tsai CL, Wei RC, Mok HK, et al. The long-term effect of listening to Mozart K.448 decreases epileptiform discharges in children with epilepsy. Epilepsy Behav. 2011;21(4):420–4.
17. Molnar-Szakacs I, Overy K. Music and mirror neurons: from motion to 'e'motion. Soc Cogn Affect Neurosci. 2006;1(3):235–41.
18. Lin LC, Chiang CT, Lee MW, Mok HK, Yang YH, Wu HC, et al. Parasympathetic activation is involved in reducing epileptiform discharges when listening to Mozart music. Clin Neurophysiol. 2013 Aug;124(8):1528–35.
19. Boso M, Politi P, Barale F, Emanuele E. Neurophysiology and neurobiology of the musical experience. Funct Neurol. 2006;21(4):187.
20. Buzsáki G, Wang XJ. Mechanisms of gamma oscillations. Annu Rev Neurosci. 2012;35(1):203–25.
21. Yizhar O, Fenno LE, Prigge M, Schneider F, Davidson TJ, O'Shea DJ, et al. Neocortical excitation/inhibition balance in information processing and social dysfunction. Nature. 2011;477(7363):171–8.
22. Iaccarino HF, Singer AC, Martorell AJ, Rudenko A, Gao F, Gillingham TZ, et al. Gamma frequency entrainment attenuates amyloid load and modifies microglia. Nature. 2016;540(7632):230–5.
23. Salimpoor VN, Benovoy M, Larcher K, Dagher A, Zatorre RJ. Anatomically distinct dopamine release during anticipation and experience of peak emotion to music. Nat Neurosci. 2011;14(2):257–62.
24. Starr MS. The role of dopamine in epilepsy. Synapse. 1996;22(2):159–94.

25. Werhahn KJ, Landvogt C, Klimpe S, Buchholz HG, Yakushev I, Siessmeier T, et al. Decreased dopamine D2/D3-receptor binding in temporal lobe epilepsy: an [18F]Fallypride PET study. Epilepsia. 2006;47(8):1392–6.
26. Rektor I, Kuba R, Brázdil M, Chrastina J. Do the basal ganglia inhibit seizure activity in temporal lobe epilepsy? Epilepsy Behav. 2012;25(1):56–9.
27. Maguire MJ. Music and epilepsy: a critical review: music and epilepsy. Epilepsia. 2012;53(6):947–61.
28. Rauscher F, Shaw G, Ky C. Music and spatial task performance. Nature. 1993;365:611.
29. Kim H, Lee MH, Chang HK, Lee TH, Lee HH, Shin MC, et al. Influence of prenatal noise and music on the spatial memory and neurogenesis in the hippocampus of developing rats. Brain and Development. 2006;28(2):109–14.
30. Hughes JR. The Mozart effect. Epilepsy Behav. 2001;2(5):396–417.
31. Shaw GL, Silverman DJ, Pearson JC. Model of cortical organization embodying a basis for a theory of information processing and memory recall. Proc Natl Acad Sci. 1985;82(8):2364–8.
32. Grylls E, Kinsky M, Baggott A, Wabnitz C, McLellan A. Study of the Mozart effect in children with epileptic electroencephalograms. Seizure. 2018;59:77–81.
33. Lin LC, Lee WT, Wu HC, Tsai CL, Wei RC, Jong YJ, et al. Mozart K.448 and epileptiform discharges: effect of ratio of lower to higher harmonics. Epilepsy Res. 2010;89(2–3):238–45.
34. Štillová K, Kiska T, Koriťáková E, Stryček O, Mekyska J, Chrastina J, et al. Mozart effect in epilepsy: why is Mozart better than Haydn? Acoustic qualities-based analysis of stereoelectroencephalography. Eur J Neurol. 2021;28(5):1463–9.
35. Quon RJ, Casey MA, Camp EJ, Meisenhelter S, Steimel SA, Song Y, et al. Musical components important for the Mozart K448 effect in epilepsy. Sci Rep. 2021;11(1):16490.
36. Rafiee M, Istasy M, Valiante TA. Music in epilepsy: predicting the effects of the unpredictable. Epilepsy Behav. 2021;122:108164.
37. Teixeira Borges AF, Irrmischer M, Brockmeier T, Smit DJA, Mansvelder HD, Linkenkaer-Hansen K. Scaling behaviour in music and cortical dynamics interplay to mediate music listening pleasure. Sci Rep. 2019;9(1):17700.
38. Feng Y. Evoked responses to note onsets and phrase boundaries in Mozart's K448. Sci Rep. 2022:14.
39. Lakatos P, Shah AS, Knuth KH, Ulbert I, Karmos G, Schroeder CE. An oscillatory hierarchy controlling neuronal excitability and stimulus processing in the auditory cortex. J Neurophysiol. 2005;94(3):1904–11.
40. Arjmand HA, Hohagen J, Paton B, Rickard NS. Emotional responses to music: shifts in frontal brain asymmetry mark periods of musical change. Front Psychol. 2017;4(8):2044.
41. Blood AJ, Zatorre RJ. Intensely pleasurable responses to music correlate with activity in brain regions implicated in reward and emotion. Proc Natl Acad Sci. 2001;98(20):11818–23.
42. Hughes JR, Fino JJ, Shaw GL. The "Mozart effect" on epileptiform activity. Clin Electroencephalogr. 1998;29(3):109–19.
43. Turner RP. The acute effect of music on interictal epileptiform discharges. Epilepsy Behav. 2004;5(5):662–8.
44. Bodner M, Turner RP, Schwacke J, Bowers C, Norment C. Reduction of seizure occurrence from exposure to auditory stimulation in individuals with neurological handicaps: a randomized controlled trial. Langguth B, editor. PLoS One. 2012;7(10):e45303.
45. Lin LC, Lee MW, Wei RC, Mok HK, Yang RC. Mozart K.448 listening decreased seizure recurrence and epileptiform discharges in children with first unprovoked seizures: a randomized controlled study. BMC Complement Altern Med. 2014;14(1) [cited 2018 Jul 25]. Available from: http://bmccomplementalternmed.biomedcentral.com/articles/10.1186/1472-6882-14-17
46. Coppola G, Toro A, Operto FF, Ferrarioli G, Pisano S, Viggiano A, et al. Mozart's music in children with drug-refractory epileptic encephalopathies. Epilepsy Behav. 2015;50:18–22.
47. Coppola G, Operto FF, Caprio F, Ferraioli G, Pisano S, Viggiano A, et al. Mozart's music in children with drug-refractory epileptic encephalopathies: comparison of two protocols. Epilepsy Behav. 2018;78:100–3.

48. Dastgheib SS, Layegh P, Sadeghi R, Foroughipur M, Shoeibi A, Gorji A. The effects of Mozart's music on interictal activity in epileptic patients: systematic review and meta-analysis of the literature. Curr Neurol Neurosci Rep. 2014;14(1):420.
49. Sesso G, Sicca F. Safe and sound: meta-analyzing the Mozart effect on epilepsy. Clin Neurophysiol. 2020;131(7):1610–20.
50. Lin LC, Lee MW, Wei RC, Mok HK, Wu HC, Tsai CL, et al. Mozart K.545 mimics Mozart K.448 in reducing epileptiform discharges in epileptic children. Evid Based Complement Alternat Med. 2012;2012:1–6.
51. McFee B, Ellis DP W. Analyzing song structure with spectral clustering. Int Soc Music Inf Retr. 2014;
52. Lahiri N, Duncan JS. The Mozart effect: encore. Epilepsy Behav. 2007;11(1):152–3.
53. Kuester G, Rios L, Ortiz A, Miranda M. Effect of music on the recovery of a patient with refractory nonconvulsive status epilepticus. Epilepsy Behav. 2010;18(4):491–3.
54. Miranda M, Kuester G, Ríos L, Basaez E, Hazard S. Refractory nonconvulsive status epilepticus responsive to music as an add-on therapy: a second case. Epilepsy Behav. 2010;19(3):539–40.
55. Rafiee M, Patel K, Groppe DM, Andrade DM, Bercovici E, Bui E, et al. Daily listening to Mozart reduces seizures in individuals with epilepsy: a randomized control study. Epilepsia Open. 2020;5(2):285–94.
56. D'Alessandro P, Giuglietti M, Baglioni A, Verdolini N, Murgia N, Piccirilli M, et al. Effects of music on seizure frequency in institutionalized subjects with severe/profound intellectual disability and drug-resistant epilepsy. Psychiatr Danub. 2017;29(3):399–404.
57. Ettinger AB, Kanner AM. Psychiatric issues in epilepsy. A practical guide to diagnosis and treatment. 2nd ed. Lippincott Williams & Wilkins; 2007.
58. McKee HR, Privitera MD. Stress as a seizure precipitant: identification, associated factors, and treatment options. Seizure. 2017;44:21–6.
59. Nakken KO, Solaas MH, Kjeldsen MJ, Friis ML, Pellock JM, Corey LA. Which seizure-precipitating factors do patients with epilepsy most frequently report? Epilepsy Behav. 2005;6(1):85–9.
60. Holmes GL. Cognitive impairment in epilepsy: the role of network abnormalities. Epileptic Disord. 2015;17(2):101–16.
61. Reisinger EL, DiIorio C. Individual, seizure-related, and psychosocial predictors of depressive symptoms among people with epilepsy over six months. Epilepsy Behav. 2009;15(2):196–201.
62. Fiest KM, Dykeman J, Patten SB, Wiebe S, Kaplan GG, Maxwell CJ, et al. Depression in epilepsy: a systematic review and meta-analysis. Neurology. 2013;80(6):590–9.
63. Raglio A, Farina E, Giovagnoli AR. Can music therapy alleviate psychological, cognitive, and behavioral impairment in epilepsy? Epilepsy Behav. 2014;31:7–8.
64. Mondanaro JF. Music therapy in the psychosocial Care of Pediatric Patients with epilepsy. Music Ther Perspect. 2008;26(2):102–9.

Music for Surgical/Perioperative Care

12

Kelly M. Webber and Myrna Mamaril

Introduction

This chapter summarizes the evidence for the use of music perioperatively (before, during and after surgery) with a focus on neurologically relevant studies, incorporating neurological surgery and neurologically relevant outcomes. Because there is a paucity of literature specifically describing the use of music for neurosurgical procedures, we extrapolate from the evidence on perioperative music-based interventions outside of neurosurgery where appropriate. We aim to place music-based interventions in an integrative milieu of non-pharmacological interventions that can enhance perioperative care and improve patient outcomes.

Self-Location

Kelly M. Webber received her graduate degrees in Music (M.Mus.); Teaching Integrated Music (MA.T); and Nursing (MSN). Salient features of her careers in education and nursing lie at the intersection of music, body, and mind. Bedside nurses are at the centre of the implementation of myriad healing modalities through education, modelling, encouragement, and the management of mental health challenges many critically ill patients face on the path to neurological and physiological recovery. Kelly M. Webber currently works in the Medical Cardiac ICU at Mayo Clinic.

K. M. Webber (✉)
Medical Cardiac ICU, Mayo Clinic, Rochester, MN, USA
e-mail: Webber.Kelly@mayo.edu

M. Mamaril
Johns Hopkins Hospital, Baltimore, MD, USA
e-mail: mmamari1@jhmi.edu

© The Author(s), under exclusive license to Springer Nature Switzerland AG 2023
K. Devlin et al. (eds.), *Music Therapy and Music-Based Interventions in Neurology*, Current Clinical Neurology,
https://doi.org/10.1007/978-3-031-47092-9_12

Dr. Myrna Mamaril's research is centred on quality and safety in acute care academic hospitals with a focus in reducing preoperative anxiety and improving pain management in surgical patients through the complementary techniques of music listening interventions for bedside nurses. Through Dr. Mamaril's clinical work with and mentorship of hospital-based nurses, she has helped build scientific inquiry capacity that inspires a culture of advocacy and innovation (in particular, using music) for improved perioperative patient care by nurses. Her work in developing nursing leaders to explore, disseminate and translate music listening studies into unit practices reflects patient advocacy for safety, advancement of the state of the science and commitment to improved professional practice.

Definitions

Clinical Terminology

Perioperative Nursing: A clinical practice nursing care continuum for all ages of patients from the preoperative phase of care to the intraoperative setting, and to the postanaesthetic/postoperative phase of care. Each setting reveals a unique nursing focus, such as the preoperative nursing tasks designed to prepare patients for their specialty surgeries.
Intraoperative: Setting in which surgical operations take place.
Postanesthesia care unit (PACU): Recovery area for patients' post- surgical interventions and a key area for the management of post-operative pain and anxiety.

Music Terminology

Minimal Music (also called *Minimalism*): A compositional technique meant to cloud tonality, reinvent harmonic structure, and challenge the auditory compass of the listener, minimalism is characterised by a steady pulse, repetition of short, accessible musical motifs (ostinato) and slow-changing harmonies. Preoperative minimalist music-listening may create a headspace that can draw a patient's focus away from their pain and anxiety related to the unknown aspects of forthcoming procedures – and potentially reduce peripheral awareness of pain [1].
Hypnotic Music: Hypnotism and music were initially connected through the work of Franz Anton Mesmer in the late eighteenth century. Though not a formal genre of musical study, the hypnotic-qualities of music have invoked trance-like states that have intrigued medical professionals for centuries. Clinical research on the "use of music as a hypnotic suggestion" demonstrated that patient-selected music can help access memories and influence emotional disposition [2, 3]. Mesmer noted that "sympathetic vibration" could be found in musical sounds, and that these could work in congruence to help reinforce focus, trance-like mental states, and invoke systemic relaxation [2].

Spotify®: A Swedish online musical database with over 80 million tracks established in 2006. Subscribers can join for free or pay a small fee for higher quality sound and no advertisements. Spotify allows listeners to choose genre, artist, mood, theme, playlists, and to share music with listeners around the world.

Preferred/Selected Music: Patients select songs and create their own playlist (e.g., from Spotify's online streaming service) rooted in their personal music preferences.

Both music-based interventions (MBI) and music therapy (MT) approaches provide a range of therapeutic benefits to mitigate preoperative stress and anxiety, as well as increase positive changes in behaviour and mood. In addition, the music modalities may reduce the surgical patients' postoperative pain perceptions through relaxation, distraction, and improved coping mechanisms [4–6]. It is important to establish the difference between MBI and MT approaches in the perioperative and acute care settings. MBI can be administered by musicians and healthcare professionals, whereas MT is often more systematic in nature and intended to promote musical experiences that foster long-term changes in the context of a therapeutic relationship; music therapy is administered by licensed music therapists [7]. At times, MT involves more interactive musical experiences whereas other MBI (such as music listening) can be more passive for the patients receiving care, particularly through the journey or preoperative, intraoperative, and postoperative care [7].

Perioperative Music-Based Interventions

Many patients report significant surgery-related anxiety and stress in the preoperative unit, which is notable given that preoperative anxiety can alter, inform, and intensify post-operative pain and medication requirements [8–11]. Surgeries often induce significantly increased levels of anxiety preoperatively, as well as compounding anxieties in the perioperative setting. As such, music listening interventions have been utilised as complementary therapies to provide comfort and stress relief, often enhancing patients' ability to rise above their surgical worries, discomforts, and preoperative pain [10, 12–17].

As anticipatory anxiety can inform pain perception, the use of music can provide a necessary distraction, providing relaxation for patients in the PACU [18]. PACU music listening interventions have demonstrated various positive physiologic outcomes, such as lowering blood pressure, heart rate and increasing oxygen saturations compared to baseline values [10, 12–17, 19–21]. In Hong Kong, Ni et al.'s clinical trial showed decreases in HR, systolic and diastolic blood pressures from baseline observed in both groups as well as reduced anxiety and improved vital signs compared with baseline values affirming both the anxiolytic effects of music on preoperative anxiety and its ability to reduce physiological indices [18–20]. Clinical trials revealed notable efficacy of the mitigation of pain and anxiety if controlled with music listening, particularly during the preoperative phase [20, 22, 23].

Postoperatively, music listening and music therapy interventions have also been shown to attenuate sympathetic nervous system (SNS) activity and increase parasympathetic nervous system (PSNS) activity, resulting in a reduction of anxiety and improved relaxation in patients [24]. Additionally, music listening has consistently been associated with improved pain and anxiety management in post-operative settings, in part due to its non-invasive approach, which provides patients with a sense of freedom and control during the surgical process [25]. The use of music listening demonstrated a significant decrease in a series of symptoms associated with preoperative anxiety including pain, nausea, shortness of breath, and feelings of depression; as well as an improvement in a generalised sense of well-being, with a notable improvement in fear and stress when the music-listening was live and not recorded [11]. The effects of listening to music for as little as 15 minutes in the preoperative phase was enough to indicate efficacy in improvement of patient anxiety in a randomized controlled trial (RCT) [16, 20].

Perioperative Music Therapy Approaches

Live music therapy is often not utilised perioperatively because the environment in many adult PACUs consists of large open rooms where patient bays are only separated by individual privacy curtains. According to Shertzer and Keck, the PACU environment is often noisy due to physiologic monitor alarms, staff and visitor conversations, and general clinical sounds making it difficult to provide soundproof rooms for patients [15]. As such, this environment is not usually conducive for optimal listening experiences for patients seeking meditative atmospheres for preoperative anxiety and postoperative pain. However, patients have reported that dimming overhead lights and using soundproof headphones while listening to music can provide distraction and a calming, comforting milieu to alleviate preoperative anxiety [20].

Live music therapy is often integrated in specialized pediatric and adult neurological populations to decrease preoperative fear and anxiety, especially using different musical instruments [26–28]. In addition to perioperative populations, live music therapy has also shown notable efficacy in the rehabilitation of patients with brain injury and chronic neurological disorders, particularly using rhythmic cadences for the improvement of motor skills and proprioception [28]. Music therapists often work with occupational and physical therapists to develop specific individualized interventional plans of care to help patients acclimate to the healthcare environment using interactive, therapeutic milieus [26–29]. Patients are often encouraged to sing along to familiar tunes engaging and helping them cope with unfamiliar surroundings [26–28]. Music is found to underscore pain management capacities and with a familiarity, found that it aids in helping patients negotiate the passing of time [30, 31]. According to Gooding, Yinger and Iocono [32], reducing the young patient's and family's preoperative anxiety through MBI can positively influence long-term healthcare outcomes through both the child's and the parents' experiences [32].

The effects of these different music modalities enhance patients' holistic well-being and coping in surgical contexts. Given the opportunity to track a single patient from the preoperative phase through surgery and into the PACU enriched the experiences of both patient and clinical caregiver, ensuring that patients' holistic needs are accurately met while receiving real time feedback on patient experiences. On a more practical level, bedside nurses could assist with any difficulties patients had with the various music-listening interventions. During the Coronavirus pandemic, many scheduled surgeries were considered "essential" and continued even when families were unable to accompany their loved ones into the PACU area(s). Therapeutic MBI can serve as a reminder that these patients are not alone and can seek solace and comfort in familiar activities or sounds that distract them from their acute anxieties regarding their status-post surgery [33].

Case Examples

Case 1: Awake craniotomy using various perioperative methods of music listening

LK is a 44-year-old female who was scheduled for an awake craniotomy for the excision and resection of a brain tumor. In speaking with her preoperative nurse, LK verbalised that she was "very frightened about the possible outcome of brain surgery and even possibly losing her speech." LK had been diagnosed just 2 weeks prior to her surgical procedure. The sudden diagnosis compounded her anxieties about having an awake craniotomy operative procedure as well as postoperative pain and potential neuro deficits. LK ranked high on the State-Trait Anxiety Inventory (STAI) questionnaire. LK's vital signs were the following: blood pressure (B/P) 169/101, heart rate (HR) 110, respiratory rate (RR) 22 breaths per minute, and SpO2 saturation 96%.

LK was told to bring her iPod the day of surgery for active music-listening using her own pre-selected music playlist, so she could listen with her noise cancelling headphones. For an uninterrupted 15 minutes, LK listened to her favourite classical music composer, Sergei Rachmaninoff. Upon assessing her disposition after the music-listening, the bedside nurse noted significant changes, finding LK in a much more relaxed state of mind. As her initial B/P was high, her nurse repeated vital signs: B/P 112/64, HR 72, RR 16, oxygen saturation 98% on room air. LK reported that she had a feeling of well-being that was statistically significant on the STAI Questionnaire ($p < 0.05$) and was "mentally prepared for surgery". Through LK's entire preprocedural and transport to the OR, there was ambient classical music playing, which helped LK cope during the surgery preparation of her skull for the awake craniotomy. Throughout her neurosurgical procedure, the music helped relax and distract her while the surgeon assessed her cranial nerves to ensure she would have no neurological deficits. In the PACU, LK reported a pain level of 5–6 on a 10-point scale. LK requested her iPod to listen to her classical music again and refused pain medication. The PACU nurse questioned LK at 15-minute intervals (15, 30, 45, 60 minutes) to rate her pain while listening to music, and – at each interval, she rated her pain "zero." LK's discharge vital signs revealed: B/P 116/74;

HR 66, RR 14–16, SpO2 saturation 98%. LK stated that the music was "Great! It really made a difference when I first came into my preoperative room, but more importantly it was wonderful when I was transported to the operating room and could hear the classical music again. It put my mind and body in a better place and immediately calmed my soul. Then in the postoperative recovery, I was able to continue listening to my music."

Case 2: Patient-selected music using Spotify®

JR is a 53-year-old male who was scheduled for a Laparoscopic Radical Prostatectomy. In speaking with his nurse, JR verbalised that he was "very worried about the outcome of his surgery and the cancer diagnosis,". JR had been diagnosed just 7 days prior to his surgical procedure. The sudden diagnosis compounded his anxieties about potential urinary incontinence, postoperative sexual dysfunction, and generalised neurological pain. JR also ranked high on the STAI questionnaire. In the preoperative unit, JR's vital signs were blood pressure (B/P) 178/92, heart rate (HR) 109, respiratory rate (RR) 20 breaths per minute, and SpO2 saturation 95%.

JR's nurse provided him with an iPod for active music-listening using Spotify, so he could choose his preferred music while listening with headphones. JR listened to his favourite country music artist for an uninterrupted 15 minutes. Upon assessing his disposition after the music-listening, the bedside nurse noticed a significant change in his disposition, finding JR in a much more relaxed state of mind. As his initial B/P was high, his nurse repeated vital signs: B/P 138/84, HR 88, RR 16, oxygen saturation 98% on room air. JR reported that he had a feeling of well-being and was "mentally prepared for surgery". In the PACU, JR reported a pain level of 6–7 on a 10-point scale. JR requested the iPod to listen to his country music again, refusing any pain medicine. The PACU nurse questioned JR at the time intervals of 15, 30, 45, and 60 minutes to rate his pain while listening to music, and he rated his pain "zero." JR's discharge vital signs revealed: B/P 122/70; HR 64, RR 16, SpO2 saturation 99%. JR stated that the music was "great! It made a lot of difference when I first came into the unit and on my stay here today. I want to listen to my music all day." JR's postoperative STAI scores were significantly reduced.

Case 3: Pre-recorded Minimalist music

LM is a 51-year-old male scheduled for a Laparoscopic Radical Prostatectomy. LM travelled from out of state and was very nervous about coming to the "big city", expressing his concern about potential critical outcomes from the procedure. This was LM's first surgery and the risks of incontinence, sexual dysfunction, and pain "were just unbearable." LM also ranked high on the preoperative STAI questionnaire. In the preoperative unit, his vital signs were blood pressure (B/P) 159/102, heart rate (HR) 99, respiratory rate (RR) 22 breaths per minute, and SpO2 saturation 93%. LM verbalised that he was "very worried about the surgical risks and the cancer diagnosis," and that these nerves had caused him to "start smoking again."

LM's nurse provided him with an iPod with a pre-recorded relaxation breathing track underscored with minimalistic music. LM was very sceptical about listening to this type of music but agreed to an uninterrupted 15-minutes of listening. LM discovered that after his listening, his breathing had slowed, and he felt more relaxed. Though LM was cautious about listening to minimalistic music, he found that it was "almost hypnotising and sedating." As his initial B/P was high, his nurse repeated his vital signs: B/P 119/72, HR 67, RR 14, SpO2 saturation 96% on room air. In the PACU, he reported to his PACU nurse that his pain was a number 9 on a 10-point scale. The nurse administered 50 mcg of Fentanyl and provided the pre-recorded relaxation breathing track underscored with minimalistic music as a complementary technique to decrease his surgical pain. After listening to the music, LM fell asleep – with vital signs revealing: B/P 118/68, HR 62, RR 16, SpO2 saturation 97%. The PACU nurse recorded LM's vital signs every 15 minutes, encouraged deep breathing, and requested updated pain scores from LM. LM's 60-minute pain score was a 5/10. LM's discharge vital signs revealed: B/P 126/69; HR 77, RR 18, SpO2 saturation 97%. LM stated that he found the meditative listening track with breathing instructions "really helpful. It seemed like I fell asleep as soon as I started listening." LM's spouse noted that he had finally slept while "listening to that different kind of music,", as he had had "insomnia since his diagnosis,". LM's postoperative STAI scores were significantly reduced.

Case Example 4: "Continuous Ambient Relaxation Environment" (C.A.R.E.)

JR was a 45-year-old male s/p bladder augmentation (augmentation cystoplasty – AC) to mitigate recurrent UTIs. JR is a non-verbal, cerebral palsy patient. In the PACU, JR went into acute respiratory failure with vital signs revealing: HR 124, B/P 82/50; RR 61, SpO2 86%. JR was persistently tachypneic and hypoxic s/p surgery and transitioned to comfort measures per his family's wishes.

CMO, or 'comfort measures only', are non-curative treatments for patient comfort when death is expected [34]. The extent to which these are standardised are often hospital-specific order sets. In this case, it was inclusive of pain control, complementary oxygen, oral secretion, and ADL support. The unit that JR was on had access to an in-room television with the C.A.R.E. channel. "Continuous Ambient Relaxation Environment "(C.A.R.E.) is a clinical-based channel that incorporates soundscapes, music, and visual atmospheres for optimising patients' sleep, reducing extraneous noises, and promotes healing [30]. JR's respiratory rate continued to be persistently >60 bpm. JR's mother was bedside and visibly distressed, as this was unresolved with CMO interventions during the day shift. At 2330, the C.A.R.E. channel was initiated per the bedside RN's suggestion. Part and parcel to Nightingale's nursing philosophy was the optimisation of environmental stimuli to help comfort patients and caregivers in stressful situations [35]. With the inclusion of ambient lighting and the C.A.R.E. channel, JR's RR slowed to 32 bpm over the course of 15 minutes of passive listening. In 60 minutes, JR's RR was 24 and, while slightly tachypneic, was no longer agonal. JR's mother verbalised her relief citing, "I cannot believe that music helped so quickly," and "He loves listening to Disney soundtracks at home and with his caregiver,". While the C.A.R.E. channel does not play Disney music, its audio tracks are instrumental, soothing, and relaxing soundscapes [30, 31]. Over the course of the night shift, JR's RR normalised at 20 bpm, and the patient was visibly more comfortable.

Conclusions and Future Considerations

In summary, discussion of perioperative MBI is best introduced during the preoperative visit to the surgeon's office or clinic where patients can choose their preferred music, stream, or upload their preferred selection of music onto their personal mobile device (or hospital-provided streaming service) before arriving on the day of surgery.

Published evidence demonstrates that both music therapy and music listening may significantly reduce preoperative anxiety and postoperative pain when compared with no MBI. However, there is a lack of evidence indicating whether this reduction of anxiety and pain corresponds directly to specific variations in the length of music listening, music therapy activities, environmental factors, compounding comorbidities, and standardized pharmacologic protocols. As such, we recommend further investigation to explore effects of music-listening as a nursing intervention to reduce anxiety in the PACU, as well as directly compare music therapy and MBI across various neurosurgical populations.

There is strong evidence to support the implementation of music listening as a nursing intervention in perioperative care to help attenuate SNS activation. It should be noted that while music listening can serve as an effective agent for preoperative anxiety and pain, there are conditions whose symptoms may be more severe than others, impacting data on the efficacy of music-listening as a mitigator to preoperative anxiety and postoperative pain for studies that involve multiple diagnoses [23, 26, 27, 30]. Additionally, patients who have a pre-existing diagnosis of anxiety and/or chronic pain may benefit from multimodal treatment in combination with MBI.

Creative options for non-pharmacologic post-operative interventions, like music listening, are part and parcel to improving patient experiences with pain and anxiety, as these alternative treatment options invoke both agency and autonomy based on personal treatment preferences [4, 6, 36].

While patients often articulate enjoyment of pre-recorded music, multiple studies indicate that self-selected music listening interventions can positively impact anxiety and pain in the PACU. Live music interventions have been well-received in neurological surgeries, particularly for pediatric populations, where several different therapeutic music modalities during the perioperative period showed beneficial postoperative outcomes [37].

Of note is the emerging evidence-based research conducted using whole body vibration and the utilization of a mean frequency derived from a pre-selected playlist manifested pre-, intra-and postoperatively (aurally and physiologically) could potentially increase the effects of activating the PSNS and improving motor and balance in patients with neurological disorders [38–40]. Though little is documented on the clinical application of vibration pre-, intra-, or post-operatively, the use of targeted temperature management, Bair huggers, warming blankets, and other bedside nursing interventions have helped improve clinical outcomes through targeted clinical practice at the bedside. With appropriate collaboration with IT programming and using pre-existing patient-selected music playlists, there is potential for development of a clinical MBI tool. This is an area that in our view deserves further clinical research and development.

References

1. McClurkin SL, Smith CD. The duration of self-selected music needed to reduce preoperative anxiety. J Perianesth Nurs. 2016;31(3):196–208.
2. Kennaway J. Musical hypnosis: sound and selfhood from mesmerism to brainwashing. Soc Hist Med. 2011;25:217–89. https://doi.org/10.1093/shm/hkr143.
3. Johnson AJ, Kekecs K, Roberts RL, Gavin R, Brown K, Elkins GR. Feasibility of music and hypnotic suggestion to manage chronic pain. Int J Clin Exp Hypn. 2017;65(4):452–65.
4. Hole J, Hirsch M, Ball E, Meads C. Music as an aid for postoperative recovery in adults: a systematic review and meta-analysis. Lancet. 2015;386:1659–71. https://doi.org/10.1016/S0140-6736(15)60169-6.
5. Walker JC, McNaughton A. Does listening to music preoperatively reduce anxiety? An evidence-based practice process for novice researchers. Br J Nurs. 2018;27(21):1250–4. https://doi-org.proxy1.library.jhu.edu/10.12968/bjon.2018.27.21.1250
6. Mishra KJ, Jesse E, Bukavina L, Sopko E, Arojo I, et al. Impact of music on postoperative pain, anxiety, and narcotic use after robotic prostatectomy: a randomized controlled trial. J Adv Pract Oncol. 2022;13:121–6. https://doi.org/10.6004/jadpro.2022.13.2.3.
7. Dahms R, Eicher C, Haesner M, Mueller-Werdan U. Influence of music therapy and music-based interventions on dementia: a pilot study. J Music Ther. 2021;58(3):12–36.
8. Ruiz Hernández C, Gómez-Urquiza JL, Pradas-Hernández L, Vargas Roman K, Suleiman-Martos N, Albendín-García L, Cañadas-De la Fuente GA. Effectiveness of nursing interventions for preoperative anxiety in adults: a systematic review with meta-analysis. J Adv Nurs. 2021; https://doi.org/10.1111/jan.14827.
9. Watts S, Leydon G, Birch B. Depression and anxiety in prostate cancer: a systematic review and meta-analysis of prevalence rates. BMJ Open. 2014;4:es03901. https://doi.org/10.1136/bmjopen-2013-003901.
10. O'Regan P, Wills T. The growth of complementary therapies: and their benefits in the perioperative setting. J Perioper Pract. 2009;19:382–6.
11. Gelatti F, Viganò C, Borsani S, Conistabile L, Bonetti L. Efficacy of live versus recorded harp music in reducing preoperative stress and fear related to minor surgery: a pilot study. Altern Ther Health Med. 2020;26(3):10–5.
12. Bojorquez GR, Jackson K, Andrews AK. Music therapy for surgical patients: approach for managing pain and anxiety. Crit Care Nurs Q. 2020;43:81–5. https://doi.org/10.1097/CNQ.0000000000000294.
13. Gillen E, Biley F, Allen D. Effects of music listening on adult patients' pre-procedural state anxiety in hospital. Int J Evid Based Health. 2008;6:24–49. https://doi.org/10.1111/j.1744-1609.2007.00097.x.
14. Ikonimidou E, Rehnstrom A, Naesh O. Effect of music on vital signs and postoperative pain. AORN J 2004;80(2):269–274, 277–78. https://doi.org/10.1016/s0001-2092(06)60564-4.
15. Shertzer KA, Keck JF. Music and the PACU environment. J Perianesth Nurs. 2001;16(2):90–102. https://doi.org/10.1053/jpan.2001.22594.
16. Wang SM, Kulkarni L, Dolev J, Kain N. Music and preoperative anxiety: a randomized control study. Anaesth Analg. 2002;94(6):1489–94. https://doi.org/10.1097/00000539-200206000-00021. PMID: 12032013.
17. Arslan S, Özer N, Özyurt F. Effects of music on preoperative anxiety in men undergoing urogenital surgery. Aust J Adv Nurs. 2007;26:46–54.

18. Ghezeljehet TN, Ardebili FM, Rafii F, Haghani H. The effects of patient-preferred music on anticipatory anxiety, post-procedural burn pain and relaxation level. Eur J Integr Med. 2017;9:141–7. https://doi.org/10.1016/j.eujim.2016.12.004.
19. Ni CH, Tsai WH, Lee LM, Kao CC, Chen YC. Minimising preoperative anxiety with music for day surgery patients: a randomised clinical trial. J Clin Nurs. 2012;21(5–6):620–5. https://doi.org/10.1111/j.1365-2702.2010.03466.x.
20. Mamaril M, Anicoche ML, Bulacan PA, Webber K, Urso S, Kaiser L. Music listening as a Postanesthesia Care Unit (PACU) nursing intervention for laparoscopic radical prostatectomy patients: a mixed method randomized comparative clinical trial. J Perianesth Nurs. 2022;37(4):e18.
21. Kelly-Hellyer E, Sigueza A, Pestritto M, Clark-Cutaia M. The analgesic properties of a music intervention in the Postanesthesia Care Unit. J Perianesth Nurs. 2023:S1089-9472(23)00018-7. https://doi.org/10.1016/j.jopan.2022.12.007.
22. Becker J. Music and trance. Leonardo Music J. 1994;4:41–51. https://doi.org/10.2307/1513180.
23. Petot T, Bouscaren N, Maillard O, Huiart L, Boukerrou M, Reynaud D. Comparing the effects of self-selected music versus predetermined music on patient anxiety prior to gynaecological surgery: a study protocol for a randomised controlled trial. Trials. 2019;20:20. https://doi.org/10.1186/s13063-018-3093-6.
24. Kavak AF, Altınsoy S, Ergil J, Arslan MT. Effect of favorite music on postoperative anxiety and pain. Anaesthesist. 2020;69:198–204. https://doi-org.proxy1.library.jhu.edu/10.1007/s00101-020-00731-8
25. Poulsen M, Coto J. Nursing music protocol and postoperative pain. Am Soc Pain Manag Nurs. 2017;19(2):172–6. https://doi.org/10.1016/j.pmn.2017.09.003.
26. Wu PY, Huang ML, Le WP, Wang C, Shih WM. Effects of music listening on anxiety and physiological responses in patients undergoing awake craniotomy. Complement Ther Med. 2017;32:56–60.
27. Jadavji-Mithani R, Venkatraghavan L, Bernstein M. Music is beneficial for awake craniotomy patients: a qualitative study. Can J Neurol Sci. 2015;42(1):7–16. https://doi.org/10.1017/cjn.2014.127.
28. Ghai S. Does music therapy improve gait after traumatic brain injury and spinal cord injury? A mini systematic review and meta-analysis. Brain Sci. 2023;13(3):522. https://doi.org/10.3390/brainsci13030522.
29. Kim SH, Choi SH. Anesthetic considerations for awake craniotomy. Anesth Pain Med. 2020;15:269–74. https://doi.org/10.17085/apm.20050.
30. Mazer S. Pain and the patient experience AHHM-40. Int J Psychol Neurosci. 2018;4(2):76–84.
31. Finlay KA, Anil K. Passing the time when in pain: investigating the role of musical valence. Psycho Neuromusicol Music Mind Brain. 2016;26(1):56–66.
32. Gooding LF, Yinger OS, Iocono J. Preoperative music therapy for pediatric ambulatory surgery patients: a retrospective case series. Music Ther Perspect. 2016;34(2):191–9. https://doi.org/10.1093/mtp/miv031.
33. Urso S, Wang J, Webber KM, Pantelyat A, Kaiser L, Anicoche ML, Bulacan T, Mamaril M. Music listening as a Postanesthesia Care Unit (PACU) nursing intervention for laparoscopic radical prostatectomy patients: a randomized comparative clinical trial. J Perianesth Nurs. 2022;37(6):848–857.e1. https://doi.org/10.1016/j.jopan.2022.01.006.
34. Dickerson SS, Khalsa SG, McBroom K, White D, Meeker MA. The meaning of comfort measures only order sets for hospital-based palliative care providers. Int J Qual Stud Health Well Being. 2022;17(1):2015058. https://doi.org/10.1080/17482631.2021.2015058.
35. Nightingale F. Notes on nursing: what it is and what it is not. First American ed. New York: D. Appleton & Co; 1860.
36. Stamenkovic DM, Rancic NK, Latas MB, Neskovic V, Rondovic GM, Wu JD. Preoperative anxiety and implications on postoperative recovery: what can we do to change our history? Minerva Anesthesiol. 2018;84:1307–17. https://doi.org/10.23736/S0375-9393.18.12520-X.

37. Brancatisano O, Baird A, Thompson WF. Why is music therapeutic for neurological disorders? The therapeutic music capacities model. Neurosci Behav Rev. 2020;112:600–15. https://doi.org/10.1016/j.neubiorev.2020.02.008.
38. Fischer M, Vialleron T, Laffaye G, et al. Long-term effects of whole-body vibration on human gait: a systematic review and meta-analysis. Front Neurol Neurosci. 2019;10:627. https://doi.org/10.3389/fneur.2019.00627.
39. Ritzmann R, Stark C, Krause A. Vibration therapy in patients with cerebral palsy: a systematic review. Neuropsychiatr Dis Treat. 2018;14:1607–25. https://doi.org/10.2147/NDT.S152543.
40. Bartel L, Mosabbir A. Possible mechanisms for the effects of sound vibration on human health. Healthcare. 2021;9:1–35. https://doi.org/10.3390/healthcare9050597.

Telehealth Music Therapy in Adult Neurological Care

13

Amy Clements-Cortés and Melissa Mercadal-Brotons

Introduction

Telehealth[1] music therapy was not widely practiced until 2020, when COVID-19 restrictions resulted in many people being isolated for extended periods and in person music therapy sessions were not possible. The pandemic forced healthcare professionals to adapt the way their jobs were performed, and music therapists were no exception. The health system was dramatically overloaded by COVID-19 and many services commonly delivered on a person-to-person basis were forced to move to a distance/virtual delivery model, including music therapy. Since that time, many music therapists are implementing telehealth innovations leading to improved access and engagement in services. Telehealth music therapy has been provided across the lifespan and has been used for individuals with neurologic disorders.

This chapter will provide an overview of telehealth in the field of music therapy and review relevant literature to highlight the efficacy of service provision with individuals diagnosed with these neurological disorders. Brief case vignettes are shared to situate the authors' experience/observations of participants and caregivers. The chapter concludes with recommendations for future clinical and research telehealth music therapy praxis.

---

[1] Also referred to as telemedicine, tele therapy, or tele practice.

A. Clements-Cortés (✉)
Music and Health Sciences, University of Toronto, Toronto, ON, Canada
e-mail: a.clements.cortes@utoronto.ca

M. Mercadal-Brotons
Escola Superior de Música de Catalunya, Barcelona, Spain
e-mail: mmercadal@esmuc.cat

© The Author(s), under exclusive license to Springer Nature Switzerland AG 2023
K. Devlin et al. (eds.), *Music Therapy and Music-Based Interventions in Neurology*, Current Clinical Neurology,
https://doi.org/10.1007/978-3-031-47092-9_13

Self-Location

I "Amy" am a credentialed music therapist (2000) and registered psychotherapist in (2015) Canada, with over 20 years experience working with individuals across the lifespan with a focus on dementia and palliative care. I am a music therapy educator, active researcher, and clinical supervisor who utilizes an integrative psychotherapy approach informed by several different music and psychotherapy practices, including Neurologic Music Therapy, the Bonny Method of Guided Imagery and Music, and Cognitive Behavioural Therapy. I have provided telehealth for adults and older adults with various issues, conducted research on telehealth and supervised students providing telehealth to persons across the lifespan including those with neurologic disorders.

I "Melissa" am a board-certified music therapist and music therapy supervisor in Spain, with over 30 years of experience working with people living with dementia and their caregivers. Currently I combine my work as educator (which includes the supervision of students), researcher and clinician specializing in older adults, with a special focus on people with dementia and their caregivers. I have participated in several projects providing music-based interventions online for people with dementia and their family caregivers. I have also participated in a research project evaluating a specific online methodology to provide music therapy interventions to people affected by dementia living at home and their principal caregivers.

Definition of Telehealth

In this chapter, we will refer to the practice of telemedicine provided by music therapists as telehealth music therapy (TMT). Telehealth is defined as the use of communication technologies to deliver health services, care and education from a distance. It can take several different forms: (a) live synchronous, implying real-time interaction between client and provider; (b) electronic transmission of documents (photos, videos, music), (c) remote client monitoring, involving the collection of client's data and transmission to a provider in another location, and (d) mobile health, including provision of public health information and targeted health care through a mobile phone [1]. A further category is the asynchronous form, which refers to communication that does not occur in real time.

Benefits and Challenges of Telehealth

A key advantage of telehealth at the height of the pandemic was continued service provision during a time of social isolation and lockdowns [2]. Additional advantages included increased access for clients in remote areas, increased opportunity for caregivers to participate in sessions with the primary client, reduced travel time and cost [2–5], and isolation reduction [6]. Those benefits have resulted in the continuation of TMT (synchronous and asynchronous) after the pandemic [7] to explore its applications with lonely older adults [6]. On the other hand, TMT is not exempt from challenges. These include issues with internet connectivity, delayed sound latency (especially with music), sound quality, digital security and privacy, difficulty making music together in real time, client familiarity with technology, and financial barriers [1, 8–12]. According to Dowson et al. [10] sessions should be adapted to accommodate technical challenges. Additional challenges include client engagement, session design, and increased support/set up needed on the part of the client with added facilitation complexity for the therapist [3].

Literature Review

Emergence of Telehealth in Music Therapy

The practice of synchronous telehealth surged during the COVID-19 pandemic. Prior to the pandemic, which imposed social distancing restrictions in all parts of the globe, TMT had not been explored in depth for persons with neurologic disorders, but was considered as an option for individuals where no music therapy services existed, including autistic children [13] (see Chap. 15), persons with spinal cord injury [14], and persons with PD who were unable to travel for music therapy [15] (see Chap. 5).

There is a growing body of literature that describes online music therapy with people with a variety of diagnoses, including neurologic disorders [10, 16]. Cole et al. [4] reported the use of NMT® techniques in sensorimotor, speech and language, and cognition domains in people ranging from children with neurodevelopmental disorders to older adults with neurologic impairments. Baker and Tamplin's [3] survey of registered music therapists found that 33.6% of therapists provided telehealth for persons with intellectual disabilities, among other broad categories such as children and adults. They noted primary reasons for continuing telehealth services as access (i.e. for rural communities and patients unable to travel to in-person sessions) and service delivery (i.e. more frequent/shorter sessions), with 43.8% of therapists reporting they would continue to provide telehealth music therapy. Clements-Cortés et al.'s [2] international study of 572 music therapists found correlations among therapist proficiency at conducting assessments via telehealth, positive benefits outweighing challenges and plans to continue telehealth use for the flexibility of scheduling, continuity of care, and sustained momentum of treatment. In addition, Vinciguerra and Federico [17] reviewed articles published between March 2020 and November 2021 to assess the feasibility of TMT for people with neurological disorders. The results show that music therapists have used many of the same techniques used for in-person interventions, with adaptations introduced to ensure the success of interventions, such as incorporating the presence of caregivers and the need for more advanced planning and structure.

While many in person music therapy sessions resumed after COVID-19 restrictions were lifted, there is still a clear need and role for telehealth. Wilhelm and Wilhelm's [18] investigation of music therapy for older adults found that 46% would continue to provide services via telehealth post pandemic. However, as a result of the COVID-19 pandemic, there has been a growing number of individuals who continue to need and choose telehealth over in-person sessions. As a result, music therapy professionals have developed innovative approaches to continue their work remotely with a range of individuals with different needs [19].

Supporting Persons with Dementia and Their Caregivers

Family caregivers play a crucial role in providing care to people with neurological disorders who live at home. In the case of dementia, care partners report that caregiving is satisfying and rewarding [20]; however, the task of managing the challenging behavioural and psychological symptoms of dementia (BPSD) can override a caregiver's capacity to cope, leading to negative physical and mental health including fatigue, depression, burnout and illness [21]. Several studies have recently focused on the use of music-based activities taught to and implemented by caregivers of persons with dementia [22–24]. For example, Bufalini et al. [25] conducted 8 music intervention sessions with 7 patient-caregiver dyads in a nursing facility. The intervention consisted of listening to personalized playlists, which facilitated increased personal connection and bonding, and a greater appreciation of the relationship among the dyads.

Baker et al. [26] designed an international study with a large sample (n = 495 dyads) of persons with dementia and their family caregivers (CGs) in their home setting to compare the implementation of a 12-week music intervention with a reading intervention by the caregivers that targeted BPSD, quality of life and well-being of PWD and CGs. The caregivers received a 2-h training by a professional music therapist prior to the intervention. The aim of the study is to demonstrate the effectiveness of the 12-week HOMESIDE music intervention in addition to standard care, on the short-term BPSD at the end of intervention of persons with dementia living at home and being cared for by a cohabiting CG.

Another study by Kim et al. [27] assessed between and within group differences in a pilot project that involved 35 participants (intervention n = 24 and no intervention or control n = 11) in an 8-week online music-based intervention for caregivers of people with dementia living at home on perceived coping, stress and depression. Post test results pointed to a significant difference between the groups ($p = 0.048$), suggesting the intervention group's coping skills were better than the comparison group. Music-based interventions also showed potential benefits to improve stress ($p = 0.028$) and symptoms of depression in dementia patients' caregivers.

In Spain, Mercadal-Brotons et al. ([28]) evaluated the effectiveness of an online program that offered personalized music resources to family caregivers and patients with Alzheimer's and other dementias who live at home under the supervision of a music therapist. The results of the Neuropsychiatric Inventory assessment showed pre-post improvements in apathy, irritability, and agitation in the people living with dementia and a decrease in the caregivers' perceived stress for each of these challenging behaviors. Caregivers also perceived a positive impact in a variety of areas through the Impact Areas Questionnaire, with the highest being facilitation of communication, offering distraction/entertainment, and providing a positive/creative experience.

However, it is important to bear in mind Baker's [29] statement that "carers cannot always be present" [during telehealth sessions]. In order to fill this void, and as an additional support to people with dementia, Baker and her team are working to create an electronic music therapy application (Music attuned technology for care via eHealth, or Match) for people with dementia to choose the best music to play to each person to alleviate agitation to be used when carers are not available [30].

TMT

Neurologic Music Therapy Approaches

Group music therapy sessions provided over Zoom for older adults with various health issues including dementia, PD and stroke can provide participants the opportunity to sing, make music together, participate in movement, and potentially afford benefits such as reduced feelings of social isolation, enhanced mood, cognitive stimulation and enhanced communication, (Personal Communication, Grace Rogers March 2022). In a reenvissioned NMT® group opportunity in Toronto, a variety of therapeutic movement experiences (TMEs) including Patterned Sensory Enhancement®, Music Attention Control Training®, Vocal Intonation Therapy®, Therapeutic Singing®, and Music and Psychosocial Counselling® were offered to participants. The group was composed of approximately 10 clients each logging in from their own homes one time per week for a 60-min session. The group was facilitated by applied music and health Master and PhD students and supervisors at the University of Toronto. Planning of the sessions was time consuming, especially at the onset, to devise the TMEs and utilize additional features with online platforms such as sharing of song lyrics and set up of web cameras to provide the clearest visuals to ensure modeling by the therapist was optimal for participants. To date, this group has chosen to continue their sessions over telehealth as it was easier for them to attend given it removed barriers such as transportation and associated financial issues [31].

Music Medicine Approaches

Vibroacoustic therapy (VAT) administered by a music therapist or a music medicine practitioner can also be monitored via telehealth. VAT is "a combination of low-frequency sound vibration, [and] music listening combined with therapeutic interaction" ([32], p. 128). Clements-Cortés worked with clients and their caregivers via telehealth prior to [33] and during the COVID-19 pandemic [34] to help them set up rhythmic sensory stimulation (RSS) treatments that caregivers could administer via the VTS1000 device in their homes. RSS incorporates vibrotactile stimulation, and auditory pulses focusing on the 20–130 Hz frequency range [35]. Clements-Cortés provided guidance via Zoom on how to set up the VTS1000 and checked in at various points to monitor device usage and answer questions. This stimulation was offered to persons with dementia and followed Clements-Cortés et al. [36], who found that RSS was successful at increasing cognition and alertness in persons with mild to moderate dementia where treatment was provided in-person. Given that live music making is not typically part of the sessions, vibracoustic therapy and RSS are quite suitable to telehealth practice, especially if there is a support person or caregiver who can assist with the sessions. There are preliminary reports of efficacy for VAT for persons with other neurological disorders including PD [37, 38]; brain injury for pain and anxiety reduction as well as spasticity [39]; and stroke [40].

Supporting Persons with PD

Group singing in individuals with PD has emerged as a viable community-based rehabilitation strategy to address voice and other impairments (see Chaps. 5 and 6). Stegemöller et al. [15] tested the feasibility of a pre-recorded Group Therapeutic Singing® program for persons with PD as a treatment for voice and respiratory impairments and found this was a feasible form of therapy as voice and respiratory outcome measures improved similar to in person sessions utilizing the same intervention. Tamplin et al.'s [41] online feasibility study utilizing therapeutic singing for persons with PD, included breathing and vocal warmup exercises, singing, conversation and social interaction to assess factors such as recruitment, retention, communication and wellbeing. Podlewska and van Wamelen [42] noted difficulties with telehealth assessments for persons with PD, including limitations in assessing motor and non-motor symptoms. This is aligned with Cole et al. [4] and Kantorová et al. [43], who reported that neurologic music therapists had a decreased use of rhythmic auditory stimulation (RAS®) for gait training via telehealth, largely owing to therapists' safety concerns (not being physically present while the client is executing the movements).

Stroke and Brain Injury

Wolfe [44] conducted a pilot study with 20 participants over 10 weeks on telehealth group music therapy sessions with persons with acquired brain injury. Results indicated statistically significant increases in self-reported well-being. Participants also reported relatively high rates of audio (78.6%) and visual quality (85.7%), adequate emotional support (85.7%), formation of a therapeutic relationship (78.6%) and connecting with others meaningfully (92.9%).

Music therapists also work with military veterans over telehealth, who have various issues such as Post-Traumatic Stress Disorder and Traumatic Brain Injury (TBI). For example, Chilton et al. [45] share a vignette of a 23-year-old female "Tina" with severe TBI who participated in music therapy weekly over a one-year period with a support worker and engaged in interventions to address communication coordination, cognitive and attention goals. Further Vaudreuil et al. [46] share examples of positive responses to music therapy via telehealth in military populations affording such benefits as decreased pain, anxiety, and depression; while participants desired to continue to engage in TMT and would recommend to peers.

Virtual Community Music Experiences

As noted above, many choirs moved their rehearsals and performances online during the pandemic. This included choirs that were formed for specific groups or needs such as for persons living with dementia, PD's and post-Stroke. 'Singing Together Measure by Measure' is one example of a music and community experience for stroke survivors and their caregivers, offered through the Mount Sinai, Louis Armstrong Department of Music Therapy New York City. The choir moved online to continue during COVID-19 restrictions. Post pandemic, this music therapist facilitated choir continues to run as a hybrid, both in person and virtually weekly for one hour and is focused on singing, sharing and reflecting on memories. (Personal communication, Joanne Loewy, August 2022[2]). A similar scenario has occurred in Baltimore, Maryland with the ParkinSonics choir of individuals with parkinsonism and their care partners (the choir began in 2016 and has been meeting online since May 2020) (see Chap. 5). Lee et al.'s [47] study of dementia-inclusive singing groups in Ireland during the pandemic suggested that these opportunities should be sustained as a means of increasing accessibility and supporting individuals with dementia living in the community.

Ethics and Telehealth

Considerable attention has surrounded platform utility and technology challenges at the start of the pandemic, with more recent focus placed on the necessity to have a balance of the needs of clients and therapists moving forward in advancing telehealth. Agres et al. [48] acknowledge that more interdisciplinary research is needed to come up with the best technology to serve each person's needs and challenges. There are many areas for additional research, including:

- how to best deliver telehealth music therapy for different age groups and for various clinical needs and potential goals [49]
- testing protocols for both music therapy and music medicine
- understanding consumer perceptions of in-person vs. telehealth experiences
- student experiences providing telehealth music therapy as done in an early study by Krout et al. [50]

---

[2] Clements-Cortés, Amy, interview by author, New York, NY, August 10, 2022.

- exploration of various aspects of the therapeutic process in music therapy, such as assessment delivered via telehealth

A key issue pertains to the music used in therapy. Music therapists often use pre-recorded music and play the music of artists. If they are not recording, there is likely a low legal risk. Where it becomes less clear is with respect to recording copyrighted music where licences and/or permissions are needed. One potential workaround is for the therapist to suggest links to publicly available recordings and resources to their clients (AMTA [51]). This then leads to another ethical issue regarding the music and the quality of the music used. Reid and Kresovich's [52] qualitative inquiry on the barriers of copyright and TMT revealed that copyright law was an obstacle that resulted in missed care opportunities given concerns music therapists had surrounding copyright liability.

Another important challenge is ensuring that patient privacy and confidentiality are maintained online. This includes ensuring that the virtual platform is secure, private, and conforms to the country's privacy laws – for example, Health Insurance Portability and Accountability Act (HIPAA) in the United States (AMTA [51]). Five of the best rated HIPAA compliant telehealth platforms include: Doxy.me, TheraLINK, VSee, SimplePractice and Zoom (Paid HIPAA-compliant version) [53]. This can become a challenge when clients are limited in their comfort and/or skills to use different platforms and request the therapist to use a non-secure platform that they are familiar with, such as Facetime.

Conclusion and Future Directions

TMT is here to stay but is not likely to replace in-person sessions for most patients with neurological disorders. Rather, it can provide complementary follow up in the clients' home and expand access to music therapy services for those unable to travel to in-person visits. TMT practice grew rapidly during the pandemic in conjunction with restrictions on in-person therapy. With a growing body of research on TMT best practices and further studies demonstrating beneficial clinical outcomes, TMT will become a fully established music therapy practice with far-reaching implications.

Crucially, the education and training of music therapists should now include telehealth practices. For example, the addition of telehealth competencies as core to the learning outcomes of a degree program is advisable in our view. It is difficult to say whether TMT placements will become a core training requirement, as it is these authors' stance that in-person sessions still need to be the focus for students developing their skills and competencies. Shenandoah University, for example, has implemented a required course for undergraduate students on telehealth and there are also several Continuing Education providers who have developed courses for professionals [54–56]. However, if TMT continues to expand, it is likely that telehealth placements will become a required practicum or aspect of internship experiences. It will be exciting to see how telehealth music therapy practice continues to grow and serve individuals with neurologic illness.

References

1. CMS.gov. Medicare telemedicine health care provider fact sheet. Published March 17, 2020. https://www.cms.gov/newsroom/fact-sheets/medicaretelemedicine-health-care-provider-fact-sheet
2. Clements-Cortés A, Pranjić M, Knott D, Mercadal-Brotons M, Fuller A, Kelly L, Selvarajah I, Vaudreuil R. International music therapists' perceptions and experiences in telehealth music therapy provision. Int J Environ Res Public Health. 2023;20:5580. https://doi.org/10.3390/ijerph20085580.
3. Baker F, Tamplin J. Music therapy service provision via telehealth in response to COVID-19 restrictions: a survey of Australian practitioners and consumers. Aust J Music Ther. 2021;32(1):1–24. https://www.austmta.org.au/public/151/files/AJMT/2021/Issue%201/3_%20AJMT%2032(1)%20-%20Baker%20%26%20Tamplin.pdf
4. Cole L, Henechowicz T, Kang K, Pranjić M, Richard N, Tian G, Hurt-Thaut C. Neurologic music therapy via telehealth: a survey of clinician experiences, trends, and recommendations during the COVID-19 pandemic. Front Neurosci. 2021;15:648489. https://doi.org/10.3389/fnins.2021.648489.
5. Mercadal-Brotons M, Pulice, G., Martí, P., de Castro, M., Vernet, M.. Online support program for the use of music-based resources for daily care for families living with dementia. Music Ther Perspect. in press
6. Kelly L, Richardson I, Moss H. Reducing rural isolation through music: telehealth music therapy for community dwelling people living with dementia and their family caregivers in rural Ireland. Rural Remote Health. 2023;23(1):8162. https://doi.org/10.22605/RRH8162.
7. Kher H, Haddad N, Zide B, Hernandez M, Hanser S, Donovan N. A personalized, telehealth music therapy intervention for lonely older adults: a feasibility study. Am J Geriatr Psychiatry. 2022;30(4 Supplement):S105–6. https://doi.org/10.1016/j.jagp.2022.01.240.
8. Clements-Cortés A, Mercadal-Brotons M, Alcântara Silva T, Moreira S. Telehealth music therapy for persons with dementia and/or caregivers. Music Med. 2021;23(3):206–10. https://doi.org/10.47513/mmd.v13i3.821.
9. Dassa A, Ray K, Clements-Cortés A. Reflections on telehealth music therapy for persons with dementia in response to COVID-19. Music Med. 2021;13(3):201–5. https://doi.org/10.47513/mmd.v13i3.818.
10. Dowson B, Atkinson R, Barnes J, Barone C, Cutts N, Donnebaum E, Hung Hsu M, Lo Coco I, John G, Meadows G, O'Neill A, Noble D, Norman G, Pfende F, Quinn P, Warren A, Watkins C, Schneider J. Digital approaches to music-making for people with dementia in response to the COVID-19 pandemic: current practice and recommendations. Front Psychol. 2021;12:625258. https://doi.org/10.3389/fpsyg.2021.625258.
11. Sasangohar F, Bradshaw M, Carlson M, Flack J, Fowler J, Freeland D, Head J, Marder K, Orme W, Weinstein B, Kolman J, Kash B, Madan A. Adapting an outpatient psychiatric clinic to telehealth during the COVID-19 pandemic: a practice perspective. J Med Internet Res. 2020;22(10):e22523. https://doi.org/10.2196/22523.
12. Sorinmade O, Kossoff L, Peisah C. COVID-19 and telehealth in older adult psychiatry-opportunities for now and the future. Int J Geriatr Psychiatry. 2020;35(12):1427–30. https://doi.org/10.1002/gps.5383.
13. Baker F, Krout R. Songwriting via skype: an online music therapy intervention to enhance social skills in an adolescent diagnosed with Asperger's syndrome. Br J Music Ther. 2009;23(2):3–14. https://doi.org/10.1177/135945750902300202.
14. Tamplin J, Loveridge B, Clarke K, Li Y, Berlowitz D. Development and feasibility testing of an online virtual reality platform for delivering therapeutic group singing interventions for people living with spinal cord injury. J Telemed Telecare. 2020;26(6):365–75. https://doi.org/10.1177/1357633X19828463.
15. Stegemöller E, Diaz K, Craig J, Brown D. The feasibility of group therapeutic singing telehealth for persons with PD's in rural Iowa. Telemed e-Health. 2020;26(1):66–70. https://doi.org/10.1089/tmj.2018.0315.

16. Garrido S, Dunne L, Stevens C, Chang E, Clements-Cortes A. Music playlists for people with dementia: trialing a guide for caregivers. J Alzheimers Dis. 2020;77(1):219–26. https://doi.org/10.3233/JAD-200457.
17. Vinciguerra C, Federico A. Neurological music therapy during the COVID-19 outbreak: updates and future challenges. Neurol Sci. 2022;43(6):3473–8. https://doi.org/10.1007/s10072-022-05957-5.
18. Wilhelm L, Wilhelm K. Telehealth music therapy services in the United States with older adults: a descriptive study. Music Ther Perspect. 2022;40(1):84–93. https://doi.org/10.1093/mtp/miab028.
19. Knott D, Block B. Virtual music therapy: developing new approaches to service delivery. Music Ther Perspect. 2020;38(2):151–6. https://doi.org/10.1093/mtp/miaa017.
20. Pione R, Spector A, Cartwright A, Stoner C. A psychometric appraisal of positive psychology outcome measures in use with carers of people living with dementia: a systematic review. Int Psychogeriatr. 2021;33(4):385–404. https://doi.org/10.1017/S1041610220003464.
21. Papastavrou E, Kalokerinou A, Papacostas S, Tsangari H, Sourtzi P. Caring for a relative with dementia: family caregiver burden. J Adv Nurs. 2007;58(5):446–57. https://doi.org/10.1111/j.1365-2648.2007.04250.x.
22. Foster S, Balmer D, Gott M, Frey R, Robinson J, Boyd M. Patient-centred care training needs of health care assistants who provide care for people with dementia. Health Soc Care Community. 2019;27(4):917–25. https://doi.org/10.1111/hsc.12709.
23. Ray K, Dassa A, Maier J, Davis R, Ogunlade O. Caring for individuals with dementia on a continuum: an interdisciplinary approach between music therapy and nursing. In: Moretti D, editor. Update on dementia. Rijeka: IntechOpen; 2016. p. 427. https://doi.org/10.5772/00000.
24. Stuart-Röhm K, Baker FA, Clark I. Training formal caregivers in the use of live music interventions during personal care with persons living with dementia: a systematic mixed studies review. Aging Ment Health. 2023; Advance online publication. https://doi.org/10.1080/13607863.2023.2180485
25. Bufalini J, Elinger P, Lehman E, Georged D. Effects of a personalized music intervention for persons with dementia and their caregivers. J Alzheimers Dis Rep. 2022;6(1):43–8. https://doi.org/10.3233/ADR-210043.
26. Baker F, Bloska J, Braat S, Bukowska A, Clark I, Hsu M, Kvamme T, Lautenschlager N, Lee YE, Smrokowska-Reichmann A, Sousa T, Stensaeth K, Tamplin J, Wosch T, Odell-Miller H. HOMESIDE: home-based family caregiver-delivered music and reading interventions for people living with dementia: protocol of a randomised controlled trial. BMJ Open. 2019;9(11):e031332. https://doi.org/10.1136/bmjopen-2019-031332.
27. Kim H, Engström G, Theorell T, Hallinder H, Emami A. In-home online music-based intervention for stress, coping, and depression among family caregivers of persons with dementia: a pilot study. Geriatr Nurs. 2022;46:137–43. https://doi.org/10.1016/j.gerinurse.2022.05.011.
28. Mercadal-Brotons M, Tomaino C, Alcantara T, Moreira S. Music therapy and music-based interventions in the context of dementia: recommendations for clinical guidelines-part II. Music Med Interdiscip J. 2021;13(3):169–73. https://doi.org/10.47513/mmd.v13i3.822.
29. Baker F. Music therapy eHealth supporting people with dementia. A music therapist and a human-computer interaction expert help carers use music to calm people with dementia. Australian Government Department of Health and Aged Care, May 26, 2022. https://www.health.gov.au/news/music-therapy-ehealth-supporting-people-with-dementia
30. Aged Care Guide. MATCH app could revolutionise in-home care for people living with dementia. 2023. Retrieved from https://www.agedcareguide.com.au/talking-aged-care/match-app-could-revolutionise-in-home-care-for-people-living-with-dementia
31. "Music and Health Applied". University of Toronto, Faculty of Music. 2022. https://music.utoronto.ca/mob-programs.php. Accessed 26 Aug 2022.
32. Punkanen M, Ala-Ruona E. Contemporary vibroacoustic therapy: perspectives on clinical practice, research, and training. Music Med. 2012;4(3):128–35.

33. Clements-Cortés A, Ahonen H, Evans M, Tang-Wai D, Freedman M, Bartel L. Can rhythmic sensory stimulation decrease cognitive decline in Alzheimer's disease? A clinical case study. Music Med. 2017;9(3):174–7. https://doi.org/10.47513/mmd.v9i3.565.
34. Clements-Cortés A, Bartel L. Long-term multi-sensory gamma stimulation of dementia patients: a case series report. Int J Environ Res Public Health. 2022;19(23):15553. https://doi.org/10.3390/ijerph192315553.
35. Bartel L, Mosabbir A. Possible mechanisms for the effects of sound vibration on human health. Healthcare. 2021;9(5):597. https://doi.org/10.3390/healthcare9050597.
36. Clements-Cortés A, Ahonen H, Evans M, Freedman M, Bartel L. Short term effects of rhythmic sensory stimulation in Alzheimer's disease: an exploratory pilot study. J Alzheimers Dis. 2016;52(2):651–60. https://doi.org/10.3233/JAD-160081. https://tspace.library.utoronto.ca/handle/1807/72411
37. King L, Almeida Q, Ahonen H. Short-term effects of vibration therapy on motor impairments in PD's disease. NeuroRehabilitation. 2009;25(4):297–306. https://doi.org/10.3233/NRE-2009-0528.
38. Mosabbir A, Almeida Q, Ahonen H. The effects of long-term 40-Hz physioacoustic vibrations on motor impairments in PD's disease: a double-blinded randomized control trial. Healthcare (Basel). 2020;8(2):113. https://doi.org/10.3390/healthcare8020113.
39. Rüütel E, Vinkel I, Eelmäe P. The effect of short-term vibroacoustic treatment on spasticity and perceived health condition of patients with spinal cord and brain injuries. Music Med. 2017;9(3):202–8. https://doi.org/10.47513/mmd.v9i3.541.
40. Jamali S, Fujioka T, Ross B. Neuromagnetic beta and gamma oscillations in the somatosensory cortex after music training in healthy older adults and a chronic stroke patient. Clin Neurophysiol. 2014;125(6):1213–22. https://doi.org/10.1016/j.clinph.2013.10.04.
41. Tamplin J, Morris M, Baker F, Sousa T, Haines S, Dunn S, Tull V, Vogel A. ParkingSong online: protocol for a telehealth feasibility a study of therapeutic group singing for people with PD. BMJ Open. 2021;11(12):e058953. https://doi.org/10.1136/bmjopen-2021-058953.
42. Podlewska A, van Wamelen D. PD's and Covid-19: the effect and use of telemedicine. Int Rev Neurobiol. 2022;165:263–81. https://doi.org/10.1016/bs.irn.2022.04.002.
43. Kantorová L, Kantor J, Horejší B, Gilboa A, Svoboda Z, Lipsky M, Mareckova J, Miloslav Klugar M. Adaptation of music therapists' practice to the outset of the COVID-19 pandemic—going virtual: a scoping review. Int J Environ Res Public Health. 2021;18(10):5138. https://doi.org/10.3390/ijerph18105138.
44. Wolfe C. Feasibility and effectiveness of group telemusic therapy with adult survivors of acquired brain injury (ABI): a retrospective pilot trial. Master's Thesis, Temple University. 2021. https://scholarshare.temple.edu/handle/20.500.12613/6840
45. Chilton G, Vaudreuil R, Freeman E, McLaughlan N, Herman J, Cozza S. Creative forces programming with military families: art therapy, dance/movement therapy, and music therapy brief vignettes. J Mil Vet Fam Health. 2021;7(3):104–13. https://doi.org/10.3138/jmvfh-2021-0031.
46. Vaudreuil R, Langston D, Magee W, Betts D, Kass S, Levy C. Implementing music therapy through telehealth: considerations for military populations. Disabil Rehabil Assist Technol. 2022;17(2):201–10. https://doi.org/10.1080/17483107.2020.1775312.
47. Lee S, O'Neill D, Moss H. Dementia-inclusive group-singing online during COVID-19: a qualitative exploration. Nord J Music Ther. 2022;31(4):308–26. https://doi.org/10.1080/08098131.2021.1963315.
48. Agres K, Schaefer R, Volk A, van Hooren S, Holzapfel A, Bella S, Müller M, de Witte M, Herremans D, Melendez R, Neerincx M, Ruiz S, Meredith D, Dimitriadis T, Magee W. Music, computing, and health: a roadmap for the current and future roles of music technology for health care and wellbeing. Music Sci. 2021;4(1):1–32. https://doi.org/10.1177/2059204321997709.
49. Richard Williams NM, Hurt-Thaut C, Thaut MH. Novel screening tool and considerations for music therapists serving autistic individuals via telehealth: qualitative results from a survey of clinicians' experiences. J Music Ther. 2023:thac009. Advance online publication. https://doi.org/10.1093/jmt/thac009

50. Krout R, Baker F, Muhlberger R. Designing, piloting, and evaluating an on-line collaborative songwriting environment and protocol using skype telecommunication technology: perceptions of music therapy student participants. Music Ther Perspect. 2010;28(1):79–85. https://doi.org/10.1093/mtp/28.1.79.
51. American Music Therapy Association. "Copyright and Licensing: Maintaining Compliance in Online/Remote Environments". Advisory from the AMTA COVID-19 Task Force, June, 2020. Accessed August 20, 2022. https://www.musictherapy.org/assets/1/7/Advisory_Copyright_from_the_AMTA_COVID19_Task_Force_06_20_20.pdf.
52. Reid A, Kresovich A. Copyright as a barrier to music therapy telehealth interventions: qualitative interview study. JMIR Form Res. 2021;5(8):e28383. https://doi.org/10.2196/28383.
53. John S. Best HIPPA compliant therapy platforms. 2023. Retrieved from https://www.counselingreviews.com/best-hipaa-compliant-therapy-platforms/
54. Music Therapy Ed. Advanced telehealth music therapy. n.d. Retrieved from https://musictherapyed.com/courses/advanced-telehealth-music-therapy/.
55. Music for Kiddos. Lights, camera, telehealth! Online course. n.d. Retrieved from https://www.musicforkiddos.com/store/lct
56. Spiegel Academy. Engaging telehealth tips and tricks for the music therapist. 2021. Retrieved from https://thespiegelacademy.com/product/engaging-telehealth-tips-and-tricks-for-the-music-therapist-2/

Therapeutic Technology for Music-Based Interventions

14

Kirsten Smayda and Brian Harris

Introduction

This chapter will explore a range of music-based therapeutic technologies intended for use in the treatment of neurological, mood, and other disorders. The technologies presented here represent a non-exhaustive spectrum of products currently being developed or utilized. By "therapeutic technologies", the authors refer to a technology that delivers an intervention or serves as a complementary tool for clinicians or patients. The spectrum of technologies presented here vary in a number of factors including user experience, clinical implementation, therapeutic intervention (e.g. music-listening, music production, coordinating movements to music); business models such as over-the-counter, subscription-based, employer-offered, and prescription-only; and the amount of peer-reviewed, published evidence that supports the product. This approach was taken to showcase the breadth of technologies being developed to deliver or complement music-based interventions (MBI) at the time of writing. The authors will take a multi-pronged approach in describing the landscape: first, the chapter will describe the relevance of technologies for MBI amongst broader trends in the healthcare system. Then, various products in the landscape of music-based technology will be introduced, followed by illustrative vignettes on how users receive and engage with two particular technologies. Given the nascency of such products and the rapid rate at which research is being done in this space, this chapter is not intended to be an exhaustive review. The authors will also highlight ongoing work across industry and academia, drawing parallels between research and commercial endeavours within the musical domain. Lastly, the authors reflect on the limitations of therapeutic technology for MBI and look

K. Smayda (✉) · B. Harris
MedRhythms, Inc, Portland, ME, USA
e-mail: ksmayda@medrhythms.com; brian@medrhythms.com

© The Author(s), under exclusive license to Springer Nature Switzerland AG 2023
K. Devlin et al. (eds.), *Music Therapy and Music-Based Interventions in Neurology*, Current Clinical Neurology,
https://doi.org/10.1007/978-3-031-47092-9_14

towards the future. The technologies discussed do not represent an endorsement by the authors of the utility or efficacy of the products.

Self-Location

Brian Harris is the CEO and Co-Founder of MedRhythms®. He is a Board-Certified Music Therapist and Neurologic Music Therapist Fellow. All of his professional training has focused on Neurologic Music Therapy (NMT®) and its clinical application across neurologic injury and disease. Brian was the first music therapist at Spaulding Rehabilitation Hospital in Boston, a teaching hospital of Harvard Medical School, and upon seeing improved clinical outcomes and a high demand for services, started MedRhythms as a way to bring NMT® to people around the world through clinical services and digital therapeutics.

Kirsten Smayda is a Senior Scientist and Medical Science Liaison at MedRhythms. Her undergraduate training was in classical singing and cognitive psychology, after which she pursued a master's degree in Music and Human Learning and a Ph.D. in Psychology from the University of Texas at Austin. During her graduate training, Kirsten studied the impact of age and music training on speech perception and sound categorization. She received a National Research Service Award fellowship, funded by the National Institutes on Aging, to study the impact of group piano lessons on speech perception in healthy older adults. Through her education, Kirsten gleaned a great appreciation for the need for evidence-based treatments to be translated into marketable options for people who need them. Her education shaped her position that stronger bridges must be built between university research activities and industry ventures to ensure that evidence-based treatments become accessible to people. Upon receipt of her Ph.D., she became a data scientist for Pear Therapeutics, a company that forged the industry of prescription digital therapeutics, before her current position at MedRhythms.

Background

In the authors' view, there has never been a better time to deliver MBI with therapeutic technology. Historically, MBI have been delivered face-to-face effectively both with and without the use of technology [1, 2], and many of these have been described throughout this book. However, there are several factors leading to why *now* is the time for technology to deliver personalized MBI to people who need it.

We live in a time when digital technology has advanced such that interventions can be delivered in a clinically meaningful way for a range of needs. Technology can now be developed with high-fidelity data collection (e.g., biometric, physiological, and kinematic), appropriate controls, feedback loops, and visual and auditory resolution – all to support the delivery of an intervention. Technology also allows for the collection and use of data throughout an intervention period in an unprecedented way. Algorithms embedded in software can utilize data streams to make informed decisions about intervention delivery specific to the user and session, which creates opportunity for appropriately challenging and progressive interventions. And with the incorporation of machine learning into interventions, robust datasets also allow for the potential to improve and apply advanced customization to the interventions themselves, over time, as well as gleaning population-level insights.

From engagement metrics to clinical patient reported outcomes (PROs) to ecological momentary assessments (EMAs); [3], data can be captured at any point in time that someone is engaging with the intervention and are important outcomes in music-based interventions [4]. Previously, the required manual effort from the clinical team or participant precluded the simultaneous collection of many data points during an intervention. Voluminous data streams can be used to refine the trajectory someone is on, report on progress to their care circle or monitor activity for safety. The data received from technologies also afford the opportunity to characterize how people change in response to music in an unparalleled context – the world outside of the lab or clinic. Technologies can travel, which allows data collection to happen in more realistic settings. The gathered evidence affords an opportunity to understand what aspects of lab-based music interventions translate best to realistic settings and how a user's response changes depending on context. However, with novel access to intervention data comes the need to make sense of the data, a trend reflected in the rising numbers of data-related positions in the United States [5]. Also, critical to this equation are the clinical and scientific perspectives needed to make sense of the rich intervention data received from the therapeutic technology of a MBI.

In addition, technology allows music to be implemented in a targeted manner, which is critical for its use therapeutically and at scale. Technology can augment, standardize, and manipulate musical input so that people perceive exactly what is intended to be played, and the musical input can be replicated to provide consistency of experience where needed. Software can selectively enhance specific features of music such as rhythm, tonal structure, dynamics, lyrics, and timbre, so that the most meaningful components of a therapeutic application are emphasized. The metadata that accompanies music such as artist, year released, tempo, groove, and genre allow for exploration and refinement of what kind of music works best therapeutically and for whom, which can help personalize treatment. Critically, technology can also incorporate other types of inputs such as wearable and non-wearable sensors, physiological monitoring, smartwatches, and smartphones to assess human health. Given the diffuse impact that music has on the brain, and the fact that the brain is shaped by sensory experience [6–8], it stands to reason that as technology evolves, so will MBI.

Some technological interventions, such as evidence-based prescription products, may represent potential alternatives or adjuncts to traditional drug treatments [9]. Of note, although technological interventions do not carry the traditional limitations of pharmaceuticals like drug-drug interactions or not as many side effects, they do come with other risks such as loss of privacy and confidentiality, and the need to securely handle large amounts of data. With such datasets, there is a need to properly address cybersecurity and develop policies and processes to appropriately handle large amounts of data.

Lastly, technology enables expanded access to important interventions and care. With a growing global age [10–12] the need for a healthcare system that can meet people where they are is paramount to offsetting the global burden of neurological disease. Technology can be shipped, packaged, shared, emailed, and texted directly to a person who needs treatment nearly anywhere around the globe, including underserved areas, without the need to manage logistics of visiting a health care setting. This phenomenon fits squarely within the macro-trends seen in the healthcare system; most notably, the rise of telehealth use during the COVID-19 pandemic (see Chap. 13). Telehealth visits grew by over 20 times since the onset of the COVID-19 pandemic for state Medicaid programs [13] without leading to an increased number of primary care visits [14], and has remained higher than pre-pandemic levels [13]. In parallel, there is growing acknowledgement of the efficacy of MBI across populations, so much so that the National Institutes of Health partnered with the University of California, San Francisco, the John F. Kennedy Center for the Performing Arts, and soprano Renée Fleming to promote research and public awareness on the topic [15].

Literature Review

Therapeutic technologies for MBI range from including music as a single component of consumer wellness products to those using music as the primary modality of intervention delivery within a prescription product. On one end of the spectrum are wellness products that capture, transmit, or store health data and are neither required to go through clinical testing nor used to treat any condition. On the other end of the spectrum are prescription digital therapeutics that are rigorously tested for clinical efficacy, regulated by the FDA, and used to treat specific conditions. For more information on the distinction, please see www.dtxalliance.org for a detailed description of this spectrum. A few decades ago, healthcare saw the emergence of teletherapy, data collection, monitoring (e.g. Medtronic, Philips Healthcare), and coaching [16] and today, emerging digital therapeutics use software to treat and manage disease. While there are many digital media through which music can be delivered for the benefit of the user's health, the focus of this chapter is on those with some peer-reviewed clinical evidence. As emphasized in the conclusion, much more peer-reviewed evidence is needed across the landscape. For readers interested in learning more about music technologies intended for use in music therapy, there are several published articles on the topic [17–20].

Two wellness products that implement music as just one component of the offering are Headspace® and Calm®. While these companies do not have published data characterizing the specific benefits of the music features they implement, they do have peer-reviewed evidence, and their size and visibility in the market warrant a description of their music components.

Headspace® is a subscription-model application, paid for by the consumer on a monthly or yearly basis, that offers mindfulness and meditation tools via video and audio tracks to help users achieve goals in areas such as sleep, movement, breath control, and work. In its "Focus" module, Headspace® offers curated music playlists created with celebrities such as Hans Zimmer, John Legend, Erykah Badu, Arcade Fire and many others, aimed at providing a musical tool to help the user's focus. In a recent systematic review, O'Daffer et al. [21] identified 14 randomized control trials (RCTs) utilizing Headspace®, although, again, no trials specifically tested the impact of the music features on any outcome [21]. Rather, the RCTs using Headspace® evaluated the effect of application use on outcomes such as general wellbeing, mindfulness, satisfaction with life, stress, symptoms of depression and anxiety, irritability, and aggression. Across the 14 RCTs, positive findings were found across mindfulness, well-being, stress, anxiety, and depression, although there are also a fewer number of studies showing negative or mixed findings for those domains. Another subscription-model application using music is Calm®, which seeks to help support sleep, meditation, and relaxation, and has a music module filled with music categorized into the categories "Mood", "Genres", "Soundscapes" and "Kids". In the same review referenced above [21], the authors identified one RCT for Calm®, where the music components were not specifically tested, and found the intervention arm resulted in improvements in stress and mindfulness scores relative to the control arm. In these two examples, the musical offerings are just one of the available activities to help users achieve wellness.

While Calm® and Headspace® utilize music as just one feature in their programs, several other products elevate music to the centerpiece of the digital experience. Some products, like Spiritune® (see Table 14.1) and Lucid, Inc., harness the power of novel, personalized music to support mood, stress, and anxiety relief and, similarly to Headspace® and Calm®, the ability to focus. Spiritune® partners with employers to deliver care to their employees (see Fig. 14.1). Although there are no published peer-reviewed papers of Spiritune® at this time, they did publish a preprint suggesting potential evidence of improved mood, using the Positive Affect and Negative Affect Scale (PANAS), and better attention, using reaction time across trials of the Flanker task. In this study, Spiritune®'s Workflow music was compared to selections from a Spotify Deep Focus playlist, Billboard's Hot 100, and an office noise control condition. This work was completed as an independent, randomized, and single-blinded large-scale study with a team at New York University's Music & Auditory Research Lab and Center for Language, Music, and Emotion.

Lucid, Inc. aims to support agitation in people with dementia, offers auditory beat stimulation, and supports Application Programming Interface integration for other products. In both offerings, the user is asked to indicate how they are feeling and also how they would like to feel. The two products then tailor a playlist to drive mental and emotional states toward the desired state. In a 2022 paper, Mallik and Russo examined the effect of Lucid's music curation system on somatic state anxiety, cognitive state anxiety, and affect in people with moderate and high trait anxiety across four auditory conditions: music, music and auditory beat stimulation (ABS), pink noise, and auditory beat stimulation alone [22]. Results suggested that participants with moderate trait anxiety in the Music & ABS condition experienced greater reductions in somatic state anxiety relative to people in the ABS-alone condition,

Table 14.1 Companies and academic labs that are developing or commercializing therapeutic technologies for MBI

Academic lab or company	Population	Goal/function area	For more information
Academic labs			
Aalborg University Copenhagen – Department of Architecture, Design, and Media Technology	Post-stroke adults	Gait impairment	spectrum.ieee.org/music-therapy#toggle-gdpr
Center for Music Technology at Georgia Tech	Adults who use prosthetics	Fine and gross motor movements	gtcmt.gatech.edu/robotic-musicianship
Center for Neurorehabilitation & Neuromotor Recovery Lab at Boston University	Parkinson's disease Stroke	Walking Gait iImpairment	bu.edu/neurorehab/ sites.bu.edu/nrl
Clinique Beau Soleil; Team Languedoc Mutualité / Nouvelles technologies	Parkinson's disease	Walking	clinicaltrials.gov/ct2/show/NCT04543058
Cognitive Brain Research Unit, Department of Psychology and Logopedics, Faculty of Medicine at University of Helsinki	Stroke Aging Aphasia Alzheimer's disease	Cognition Emotion Verbal expression Memory Self-awareness	helsinki.fi/en/researchgroups/cognitive-brain-research-unit/research/ongoing-projects
David Ott Lab for Music & Health at Montclair State University	Adults	Community support Pain support Expression of selfhood Agency	montclair.edu/john-j-cali-school-of--music/david-ott-lab/
MIT Media Lab – Oniorb Project	Adults	Stress	media.mit.edu/projects/oniorb/overview/
Music and Audio Research Laboratory & Center for Language, Music and Emotion at New York University	Stroke Aging	Cognition Motor function Mood	steinhardt.nyu.edu/marl clame.nyu.edu/
Music Brain Health Technology Lab at Leiden University	Stroke	Upper limb mobility	rebeccaschaefer.net/
SMART Lab at Toronto Metropolitan University	Adults	Anxiety	sites.psychlabs.ryerson.ca/smartlab/
The Center for Health Technology at University of Plymouth - RadioMe Project	People with dementia who live independently	Independent living	plymouth.ac.uk/research/centre-for-health-technology/radio-me

(continued)

Table 14.1 (continued)

Academic lab or company	Population	Goal/function area	For more information
Companies			
Brain.FM™	General population	Attention	brain.fm
Lucid, Inc.	Alzheimer's disease and dementia Parkinson's disease Adults Children	Anxiety Pediatric stress Burnout Agitation Depression Pain Tinnitus	lucidtherapeutics.com
MedRhythms®	Stroke Multiple sclerosis Parkinson's disease Cognitive impairment	Brain health Walking	medrhythms.com intandemrx.com
Music Health Technologies, SingFit	Cognitive decline Alzheimer's disease and dementia Caregivers Long-term care staff Rehabilitation therapists	Mood Engagement in activities Autobiographical memory	singfit.com/
Spiritune®	Adults	Promoting desired emotional state	spiritune.com
Vera™ VeraPro®	People with dementia Caregivers of people with dementia	Quality of Life Mood	veramusic.com

Fig. 14.1 Example flow demonstrating access to Spiritune®

and greater reduction in cognitive state anxiety relative to people in any other condition. Participants with high trait anxiety experienced a greater improvement in positive affect during the combined music and ABS condition relative to people in the pink noise condition, and greater improvement in negative affect in the combined music and ABS condition relative to people in the ABS-alone condition. Interestingly, the authors also reported that people with high trait anxiety in the music-only condition showed greater improvements in both somatic and cognitive state anxiety relative to people in the ABS-only condition.

Vera™, created by Music Health, is another music-based product intended for people living with dementia; however, it is designed to be implemented with the patient by the caregiver. The offerings are marketed in two forms: Vera™, for families and primary care partners of individuals living with dementia; and VeraPro®, for caretaking communities to administer music to multiple people. Vera™ and VeraPro® are available in the UK, US, Canada, Australia, and New Zealand, and also through Walgreens Find Care. Caregivers can create profiles for their clients, assess the mood of their client, and Vera™ will then curate a playlist for the person they are providing care for. There are three contexts in which a user can select music, depending on the intention of the listening session: relax, reminiscence, and energize. Vera™ also allows a user to tailor the music based on the listener's musical preference and age, so that the listening experience can be personalized to the individual. Whereas Spiritune® and Lucid generate their own music, Vera™ utilizes existing music from Universal Music Group's catalog. When the authors reached out to Vera™ to ask about published research, a team-member indicated that although there is no published research, they are engaged in research with Prince of Wales hospital in Australia.

At SingFit, creators have developed digital, musical offerings specifically for caregivers, rehabilitation therapists, and staff in long-term care settings to use with loved ones and patients. For caregivers, the SingFit STUDIO caregiver offering includes an application that has filters for age-based playlists, and also includes a feature that auditorily prompts the users with song lyrics. For rehabilitation therapists, SingFit's STUDIO Pro offers a musical tool for those working with older adults with cognitive decline to sing together. Lastly, SingFit PRIME is a singing program for use by the staff at long-term care facilities, allowing them to engage with their residents and create singing opportunities. A study using SingFit compared three conditions across 5 weeks: a "Sing" condition in which participants received the SingFit-R application and were asked to learn and sing 12 songs throughout the protocol; a "Listen" condition in which participants received the SingFit-R application and were asked to listen but not sing to the same amount of music; and a passive control condition in which participants were not asked to sing or listen to music. The authors found that participants in the "Listen" and "Sing" conditions had a trend-level improvement in visuospatial working memory, and trend-level group difference on a pre-to-post emotional regulation metric. The non-conformance rate and attrition rate were 18% and 12%, respectively. The authors conclude that given the self-motivation required to use the product, study duration, and sample size, a larger study is warranted [23].

14 Therapeutic Technology for Music-Based Interventions

MedRhythms is another company harnessing the power of music to offer the world's first prescription music product (see author locations above). In 2016, MedRhythms launched its digital therapeutics division to focus on building evidence-based technology designed to deliver interventions at home. The platform leverages auditory-motor entrainment in which an external rhythmic stimulus, like music, is presented, which causes the brain and motoric actions to naturally synchronize to the stimulation. Auditory-motor entrainment is foundational to Rhythmic Auditory Stimulation, an NMT® intervention (see Chap. 5). MedRhythms has since completed several clinical trials across neurologic diagnoses and has built the world's first prescription music platform that addresses walking challenges due to brain injury or disease using popular music sourced from a major record label. The most advanced asset in the platform is InTandem™, a medical device indicated to improve chronic stroke walking impairment and ambulation in the home and carries robust clinical data [24–27]. In Tandem is commercially available through a prescription from a healthcare professionals and may be reimbursed by health insurance. Each song included in the product undergoes a proprietary screening process to ensure therapeutic suitability. The products are built to be used in the home autonomously without clinician or caregiver input to the device (see Fig. 14.2). Users receive a kit in the mail that includes a locked touchscreen device loaded with the software, a headset, two 3D motion foot sensors, and a charging hub. As the user walks, the sensors capture components of their walking that are used as inputs to the application. The proprietary algorithm then modifies the music delivered through the headset to produce clinically-relevant walking outcomes.

In a feasibility study, a single session of the MedRhythms platform improved usual and fast-walking speed in chronic stroke survivors [24]. Further, it was found that a single session of walking with the platform could improve both asymmetry and walking efficiency in people with chronic stroke walking challenges [25]. In addition, MedRhythms conducted a pivotal trial in people with chronic stroke who experience walking challenges as a result of their stroke and demonstrated an improvement in walking characteristics such as gait speed and ambulation status, relative to an active walking control [27]. Other products in the pipeline include walking-based products for people with Parkinson's disease and multiple sclerosis. A feasibility and proof-of-concept study in people with Parkinson disease met its primary endpoint of adherence in a protocol of use five times a week for 4 weeks

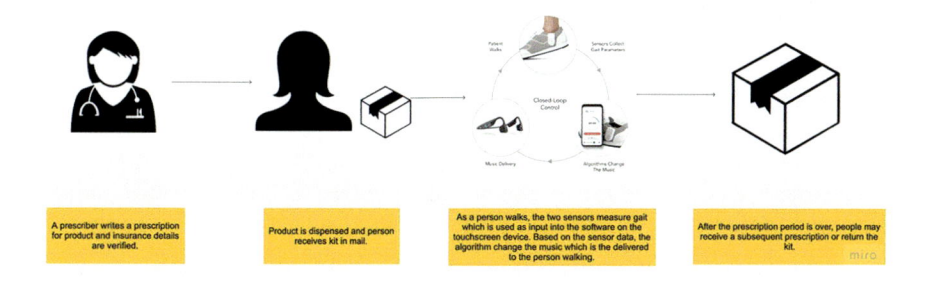

Fig. 14.2 Proposed flow for access to and use of MedRhythms products

[28]. In addition, the intervention was able to support walking at moderate intensity, and indicated positive clinical signals of improvement on quality of life, mobility, walking endurance and disease severity.

Case Examples

As MBI are developed and marketed, it can be helpful to consider the different ways in which someone might receive and use the product. Presented here are two case studies regarding Spiritune® and MedRhythms, as examples for how people might engage with technology offering MBI.

Case 1: There are two routes through which an interested person can access Spiritune®: through an employer-sponsored offering or through a direct-to-consumer approach with a monthly subscription rate. Someone who is interested in using Spiritune® can download the application to their mobile device and authenticate their account using an access code to access content through the app. At the start of a typical session, the user inputs three pieces of information that allows the product to tailor the session: (1) How are you currently feeling? (2) How would you like to feel? And (3), the context in which the person is using the product at that moment: Waking Up, Workflow, Winding Down, and Sleep. Depending on the user's input, the product will then curate a series of musical pieces with unique transitions that move the person from where they are to where they want to be, emotionally. The music presented in Spiritune undergoes a unique process that uses input from artists, neuroscientists, and machine learning to produce a library capable of delivering an intended effect. More specifically, a team of neuroscientists develop "sonic recipes," which are combinations of musical attributes that are mapped to emotional and cognitive states. Then, musical artists compose individual tracks based on the sonic recipe with their own creative freedom. Lastly, music is reviewed by both a human and a machine learning algorithm to ensure the music delivers the recipe precisely as is intended. Sessions are 10 min long, and at the end of each session a survey question asks, "Did you feel like you reached your desired response?" Amongst ~4500 users who responded, 91% answered "yes" (Spiritune® data on file).

Case 2: During the prescription and order process for MedRhythms products, it is envisioned that a physician would determine the appropriate treatment period for the patient. Once the product is dispensed and received by the individual, they are guided through onboarding with provided information. Each session lasts approximately 30 min and at the end of each session the individual will have the option to provide a response to a survey question about their experience. At the end of the intervention period the patient will return the kit.

Presented in Table 14.1 is information and resources for the products described above and additional companies and academic labs that are engaged in developing therapeutic technologies for MBI. The authors hope that this summary illustrates opportunities for collaboration, innovation, and community across academic and corporate endeavors in the music technology intervention domain.

Conclusion and Future Considerations

This chapter provided an overview of existing therapeutic technologies for MBI. Although this overview is non-exhaustive, the products presented here include a range of intervention experiences, business models, and amount of peer-reviewed evidence in order to share the breadth of the industry's offerings at the time of writing. Advancements in research on music and health have demonstrated the importance for its implementation in treatment of neurologic disorders [4]. Technology can deliver advanced and efficacious musical treatment, capture and use data for a variety of purposes, is a timely alternative to pharmacological treatment options, and can scale care to people in need globally.

To further develop MBI technologies, it is imperative that more research is conducted and published through the peer-review process. Critically, research on implementation of MBI can help pave the way to effective use in real-world settings, and research on health economics will help build the value case for music use in healthcare systems. As music-based technology interventions become more prevalent, achieving a healthy balance between clinical trial rigor and realistic implementation is both difficult and critical in developing the evidence base. This is crucial for acceptance by both the academic and commercial communities [29] but poses a serious challenge since music allows for creativity and highly individualistic behavior within a constrained environment [30].

When imagining a future with therapeutic technologies delivering MBI, a key need is more avenues for access. Without equitable routes of access, the gap in equitable healthcare will only continue to grow. Pathways such as prescription-based products that seek reimbursement from health insurance companies may be one way to alleviate the cost burden on the individual and also bring music-based treatments into standard of care. This route of access is becoming standard in countries like Germany, where, at time of writing, over 40 digital health applications are currently available via prescription on a permanent or provisional basis [31]. However, concerns continue to exist around technology engagement and equity and its role within digital health technologies [27]. In the United States, there are only two prescription digital therapeutic products covered by several state Medicaid programs and both are for treating addiction: reSET for substance abuse and reSET-O for opioid use. Three states categorize these two products as covered benefits under Medicaid, and eight states cover them under other state mechanisms [31]. The Access to Prescription Digital Therapeutics Act of 2023, which would require the Centers for Medicare and Medicaid to create a Medicare billing pathway, holds promise for wider adoption and reimbursement of prescription digital therapeutics; however, as of June 2023 the act has not passed the Senate.

In addition to carving out a feasible financial pathway to music-based digital interventions, it is critical to maintain a clinical culture within the digital product development cycle. In order for MBI to make good on their promise, accredited music therapists must be part of the development process. Without the input of clinicians, digital MBI risk missing best clinical practices, may not be implemented appropriately, and product claims may drift from the scientific evidence base that supports the therapeutic outcome. This is why it is critical for all parties to work together: music therapists, academic researchers, product developers, and commercial partners.

Although the offerings explored in this chapter represent progress in bringing therapeutic technologies for MBI to the public, they have limitations. It is likely that not every component of the evidence-based practices of music therapists is appropriate for translation into a digital medium. For instance, some components of a music therapy session in which a therapist records data about how the client performs may be made redundant from the inherent data collection that occurs during a session with an implemented technology. The rapport that is built between a therapist and a client, which is a key ingredient in therapy practice, may also be difficult to replicate, although it is possible that a different kind of relationship with the therapeutic technology and clinician could develop. Additionally, there may be certain clinical presentations or diagnostic severity levels that the current state of technology is not equipped to address. Lastly, the perception among the general population of music's role in supporting a healthier life, and as a medicinal solution, is still growing and will continue to grow as the evidence-based products described in this chapter take root across the world. Despite these challenges, the future for MBI technology is bright.

References

1. Viega M. A humanistic understanding of the use of digital technology in therapeutic songwriting. Music Ther Perspect. 2018;36(2):152–60.
2. Thaut M. Rhythm, music, and the brain: scientific foundations and clinical applications. New York: Routledge; 2007. 272 p.
3. Shiffman S, Stone AA, Hufford MR. Ecological momentary assessment. Annu Rev Clin Psychol. 2008;4:1–32.
4. Edwards E, St Hillaire-Clarke C, Frankowski DW, Finkelstein R, Cheever T, Chen WG, et al. NIH music-based intervention toolkit: music-based interventions for brain disorders of aging. Neurology. 2023; https://doi.org/10.1212/WNL.0000000000206797.
5. Occupational Outlook Handbook, Data Scientists [Internet]. Bureau of Labor Statistics, U.S. Department of Labor. 2023 [cited 2023 Jun 1]. Available from: https://www.bls.gov/ooh/math/data-scientists.htm
6. Barrett KC, Ashley R, Strait DL, Kraus N. Art and science: how musical training shapes the brain. Front Psychol. 2013;4:713.
7. Manini B, Vinogradova V, Woll B, Cameron D, Eimer M, Cardin V. Sensory experience modulates the reorganization of auditory regions for executive processing. Brain. 2022;145(10):3698–710.
8. Skoe E, Kraus N. A little goes a long way: how the adult brain is shaped by musical training in childhood. J Neurosci. 2012;32(34):11507–10.
9. Tang SK, Tse MMY, Leung SF, Fotis T. The effectiveness, suitability, and sustainability of non-pharmacological methods of managing pain in community-dwelling older adults: a systematic review. BMC Public Health. 2019;19(1):1488.
10. Ageing and health [Internet]. 2022 [cited 2022 Dec 26]. Available from: https://www.who.int/news-room/fact-sheets/detail/ageing-and-health

11. Mozafaripour S. The 2021 American nursing shortage: a data study [Internet]. University of St. Augustine for Health Sciences. 2021 [cited 2022 Dec 26]. Available from: https://www.usa.edu/blog/nursing-shortage/
12. Rosseter R. Fact sheet: nursing shortage [Internet]. American Association of Colleges of Nursing. 2022 [cited 2022 Dec 26]. Available from: https://www.aacnnursing.org/Portals/42/News/Factsheets/Nursing-Shortage-Factsheet.pdf
13. Chu RC, Peters C, Lew ND, Sommers BD. State medicaid telehealth policies before and during the COVID-19 public health emergency. Health Serv Res. 2021;58(5):988–98.
14. Dixit RA, Ratwani RM, Bishop JA, Schulman K, Sharp C, Palakanis K, et al. The impact of expanded telehealth availability on primary care utilization. Npj Digital Med. 2022;5(1):1–3.
15. Sound Health Network [Internet]. Sound Health Network. 2022 [cited 2022 Dec 28]. Available from: https://soundhealth.ucsf.edu/home
16. Jimison H, Shapiro M, Pavel M. Adaptive health coaching technology for tailored interventions. Int J Environ Res Public Health. 2021;18(5):2761.
17. Werger C, Groothuis M, Jaschke AC. Music-based therapeutic interventions 1.0 from music therapy to integrated music technology: a narrative review. Music Med. 2020;12(2):73–83.
18. Johnston D, Egermann H, Kearney G. Innovative computer technology in music-based interventions for individuals with autism moving beyond traditional interactive music therapy techniques. Särkämö T, editor. Cogent Psychol. 2018;5(1):1554773.
19. Schneider L, Gossé L, Montgomery M, Wehmeier M, Villringer A, Fritz TH. Components of active music interventions in therapeutic settings—present and future applications. Brain Sci. 2022;12(5):622.
20. Ragone G, Good J, Howland K. How technology applied to music-therapy and sound-based activities addresses motor and social skills in autistic children. Multimodal Technol Interact. 2021;5(3):11.
21. O'Daffer A, Colt SF, Wasil AR, Lau N. Efficacy and conflicts of interest in randomized controlled trials evaluating headspace and calm apps: systematic review. JMIR Ment Health. 2022;9(9):e40924.
22. Mallik A, Russo FA. The effects of music & auditory beat stimulation on anxiety: a randomized clinical trial. PLoS One. 2022;17(3):e0259312.
23. Reid AG, Rakhilin M, Urry HL, Thomas AK, Patel AD. New technology for studying the impact of regular singing and song learning on cognitive function in older adults: a feasibility study. Psychomusicology. 2017;27(2):132–44.
24. Hutchinson K, Sloutsky R, Collimore A, et al. A music-based digital therapeutic: proof-of-concept automation of a progressive and individualized rhythm-based walking training program after stroke. Neurorehabil Neural Repair. 2020;34(11):986–96. https://doi.org/10.1177/1545968320961114.
25. Collimore AN, Roto Cataldo AV, Aiello AJ, et al. Autonomous control of music to retrain walking after stroke. Neurorehabil Neural Repair. Published online June 5, 2023:154596832311742. https://doi.org/10.1177/15459683231174223.
26. Smayda KE, Cooper SH, Leyden K, Ulaszek J, Ferko N, Dobrin A. Validating the safe and effective use of a neurorehabilitation system (InTandem) to improve walking in the chronic stroke population: Usability Study. JMIR Rehabil Assist Technol 2023;10:e50438. URL: https://rehab.jmir.org/2023/1/e50438/. https://doi.org/10.2196/50438.
27. Awad L, Jayaraman A, Nolan K, et al. Efficacy and safety of using auditory-motor entrainment to improve walking after stroke: A Multi-Site Randomized Controlled Trial of the InTandem™ Neurorehabilitation System. Pre-Priint; 2023. https://doi.org/10.21203/rs.3.rs-3031294/v1.
28. Zajac JA, Porciuncula F, Cavanaugh JT, et al. Feasibility and proof-of-concept of delivering an autonomous music-based digital walking intervention to persons with parkinson's disease in a naturalistic setting. JPD. 2023;13(7):1253–65. https://doi.org/10.3233/JPD-230169.
29. Habibi A, Kreutz G, Russo F, Tervaniemi M. Music-based interventions in community settings: navigating the tension between rigor and ecological validity. Ann N Y Acad Sci. 2022;1518(1):47–57.

30. Laying the foundation: defining the building blocks of music-based interventions [Internet]. NCCIH. 2022. [cited 2022 Dec 26]. Available from: https://www.nccih.nih.gov/news/events/laying-the-foundation-defining-the-building-blocks-of-music-based-interventions
31. PalmHealthCo. Global DTx reimbursement landscape [Internet]. LinkedIn. 2023 [cited 2023 Jun 9]. Available from: https://www.linkedin.com/posts/michaelpace-dtx_dtx-diga-sleep-activity-7010953082591191040-8I7L/?utm_source=share&utm_medium=member_desktop

Music Therapy and Music-Based Approaches with Autistic People: A Neurodiversity Paradigm-Informed Perspective

15

Hilary Davies and Michael B. Bakan

Introduction

This chapter is written from the perspective of the neurodiversity paradigm, which views autism not as a disorder, but rather as a neurological difference that is a natural part of human diversity [1, 2]. We aim to raise a challenge to the concept of interventions in music therapy and other forms of music-based work with autistic people. (Note: In alignment with the preferences of most autistic people [3] we use identity-first language [e.g., autistic person] rather than person-first language [e.g., person with autism] in this chapter.)

The idea of an intervention sets up a particularly uneven power dynamic between therapist (or musician) and autistic person, with the implication that the therapist "knows best", and that the behaviour or personality of the autistic person is so flawed or disordered that it must be strongly dealt with. Instead, we promote working *with* autistic people over working *on* them, and collaboration over intervention (as Devlin puts it, working as "co-pilot" rather than "chauffeur") [4]. The traditional "medical model" [5]/"consensus model" [6] approach to music therapy with autistic people, which centres on the professional expertise of the therapist in a strictly boundaried, hierarchical, and change-focused relationship, has been challenged in recent years by approaches such as anti-oppressive music therapy that emphasise the responsibility of the therapist to examine and challenge the power dynamics and dominant societal narratives which may, consciously or unconsciously, be part of a therapeutic process [7, 8]. Similarly, ethnomusicologists who work with neurodivergent populations proceed from a foundational premise that existing power

---

H. Davies (✉)
Guildhall School of Music and Drama, London, UK

M. B. Bakan
Florida State University, College of Music, Tallahassee, FL, USA
e-mail: mbakan@fsu.edu

dynamics and cultural narratives are and must always be subject to vigorous critique and redress. Autistic people occupy a marginalised place in society as an oppressed, stigmatised neurominority (neurological minority) [3, 9], yet it can be argued that allistic (non-autistic) identities are not objectively "better" than autistic ones: as Nick Walker states, "to describe autism as a disorder represents a value judgement rather than a scientific fact." [3] p. 26).

A neurodiversity paradigm-informed approach to music therapy (which is growing in prominence [10–14]) maintains the basic principles of a therapeutic process leading to therapeutic outcomes, but rather than focusing on implementing interventions to ameliorate deficits, it instead seeks to work with autistic clients to identify goals that enhance their "strengths, confidence, self-acceptance and coping strategies" [11]. This approach can be seen to share characteristics with the approaches of ethnomusicologists working at the intersection of music and neurodiversity [15–18].

Self-Location

Despite our different professional backgrounds, we as authors share an individualised, person-centred approach to work with autistic people: we are not focused on addressing or seeking to alter the inherent neurological difference that is autism, but rather on supporting the individual with their own particular strengths and challenges. We seek to privilege the person over the "disorder", emphasising the creation of a collaborative musical relationship rather than a "musical intervention", or indeed any form of remediation or cure.

We are:

1. A UK-based, autistic music therapist (Davies) who specialises in working with autistic people, and who is currently pursuing a PhD in Music Therapy with that research focus.
2. A US-based ethnomusicologist (Bakan) whose work focuses on the musical lives of autistic people, both through ethnography and shared musical experience.

It is important to acknowledge that a fundamental difference in professional *responsibility* between the music therapist and ethnomusicologist distinguishes both the opportunities and constraints of our respective approaches. The history of the regulation of music therapy as a health profession (music therapists in the UK are regulated by the Heath and Care Professions Council [19]), as well as the expectations of the individual or organisation seeking the expertise of a therapist, can place a certain pressure or responsibility on the music therapist to produce evidence of change through the therapeutic process.

The ethnomusicologist, contrastingly, is not generally charged with the responsibility of being an agent of change. As Bakan puts it, the subjects of his musicological and applied work are *"experts at being who they are,* whether musically, culturally, socially, or in any other way," and his job as an ethnomusicologist is "not to endeavor to change them, but rather to try to understand and relate to them, to the greatest extent possible… 'on their own terms' [15].

Definitions: From Pathology to Neurodiversity

The term *neurodiversity* first officially appeared in Judy Singer's 1998 thesis [1], which introduced the concept of "neurologically different" people (such as autistic people) as a "new addition to the familiar political categories of class/gender/race" [1]. The concept of "neurodiversity" includes all kinds of human brains, while "neurodivergent" refers to brains which differ from the neurological majority (i.e. the type of brain that most humans have), described as "neurotypical" [3].

The autistic writer Nick Walker has called for a paradigm shift, from a pathology paradigm of autism to a neurodiversity paradigm [2]. The pathology paradigm is exemplified by the current diagnostic criteria for autism in the *Diagnostic and Statistical Manual of Mental Disorders 5* (DSM 5), which are applied in the majority of autism diagnoses today [20]. The pathology paradigm (related to the "medical model" of disability [5, 21]) privileges the idea that there is one "right", "normal" or "healthy" kind of human brain, and "something wrong" with neurological configurations that differ from this [2]. Music therapy has traditionally adhered to medical model narratives around disability in order to establish itself as a professional discipline [22]. Pathology paradigm/medical model approaches to music therapy might ask the question: "What do we do about the problem of these people not being normal?" [3] p. 26.

The neurodiversity paradigm, eschewing the concept of a "normal" brain [2], contrastingly views neurological diversity as a natural, healthy and valuable form of human diversity. Since autistic people are viewed within the neurodiversity paradigm as a minoritized identity (neurominority), neurodiversity paradigm-informed music therapists might ask questions including, "What do we do about the problem of these people being oppressed, marginalized, and/or poorly served and poorly accommodated by the prevailing culture?" ([3], p. 26).

The behavioural observations of the DSM 5 criteria can be seen as inadequate to define the fundamental neurological difference of the autistic mind, which leads to an inherently different experience and perception of life. As Sinclair puts it:

> Autism isn't something a person *has*, or a 'shell' that a person is trapped inside. There's no normal child hidden behind the autism. Autism is a way of being. It is *pervasive*; it colors every experience, every sensation, perception, thought, emotion, and encounter, every aspect of existence. ([23], p. 1)

Literature Review

Music Therapy Approaches

Autism and Music Therapy: Origins and Overview

Music therapists have worked with autistic people since the earliest days of professional music therapy practice [24, 25]. There have been a great variety of approaches and methods, including psychoanalytic approaches [26], music-based approaches [25] and behavioural approaches [27]. A range of contemporary approaches are covered in *Music Therapy and Autism across the Lifespan: A Spectrum of Approaches* [28].

While behavioural approaches to music therapy often use structured musical activities such as pre-composed songs focused on training in specific skills (e.g., social skills), improvisational music therapy is not usually pre-structured: the musical content is developed spontaneously by the music therapist in response to the client. Behavioural approaches are more common in the USA [29], whilst improvisational approaches have tended to be more widely used in Europe [30].

Behavioural Music Therapy: A Contested Approach

Behavioural approaches to therapy with autistic people, such as Applied Behavioural Analysis (ABA), have been strongly criticised by autistic adults who experienced them as children [31]. ABA has the stated goal of making autistic children "indistinguishable from their peer [32] (i.e., appear neurotypical). This is very much at odds with the neurodiversity paradigm, which values autistic identity. Behavioural approaches to music therapy include any approach that may seek to directly or indirectly change the behaviour of an autistic person, and can therefore be considered very much in line with the pathology paradigm.

In a 2013 study, 54.2% of music therapists in the USA described using behavioural approaches with autistic clients [29]. Martin details a number of behavioural concepts that she considers to be essential knowledge for music therapists yet offers little in the way of a rationale for making interventions based on these concepts, let alone clear evidence of how their application might actually benefit autistic children [27].

Within UK music therapy, overtly behavioural approaches are less common, and yet music therapists may find themselves working towards behavioural goals due to the preferences or requests of other professionals working with the child, such as teachers. The large-scale TIME-A trial of music therapy (The Trial of Improvisational Music Therapy's Effectiveness for Children With Autism) implicitly endorsed behavioural approaches by measuring differences in behaviour (symptom severity) according to the ADOS (Autism Diagnostic Observation Schedule) assessment tool, although many of the music therapists involved in the study may not actually have been using behavioural methods [33].

Behavioural methods of therapy can be psychologically harmful because they do not actually make people less autistic, but instead encourage autistic masking (imitating the behaviour of allistic [non-autistic] people in an attempt to cover up autistic identity), which can lead to exhaustion, burnout and mental health issues.

Community Music Therapy: A Step Towards Diversity

Community music therapy is defined by Stige as "music therapy *in* a community context" or "music therapy *for change* in a community" [34]. It takes an ecological approach, considering the social and cultural context of each client, which can facilitate taking account of the issues facing autistic people as an oppressed neurominority.

Shiloh and Lagrasse's 2014 article "Sensory Friendly Concerts (SFC)" [14] was one of the earliest music therapy publications to consider the concept of neurodiversity. Sensory friendly concerts, or SFCs, provide "accommodations for persons with different sensory needs" and an environment where "different responses to music are respected" ([14], p. 118). Although SFCs are concerts rather than therapy sessions, the therapeutic intention is to provide a deeply accepting environment where autistic audience members/participants can behave in a way that is entirely natural and comfortable: moving about the space, interacting with the performers and even taking part in the performance. SFCs address Stige's concept of "therapy *for change*" [34] by aiming to increase understanding and acceptance of autistic culture and promote social justice for autistic individuals.

Other neurodiversity paradigm-informed community music therapy projects include Young's study of singing groups for autistic adults [35]. The emphasis in this study is on "enhancing various QoL [quality of life] domains" rather than "symptom-focused interventions" ([35] p. 10).

Neurodiversity and Music Therapy

Although the neurodiversity paradigm has yielded an established narrative within autism studies spanning more than 20 years [1, 23, 36], it is only within the last few years that this perspective has been seriously addressed within the music therapy literature [10–13].

Several music therapists have written about the development of a neurodiversity paradigm-informed approach in principle [10–12], yet there remains a dearth of published case studies or clinical research in this area. A notable exception is Pickard's chapter on "Valuing Neurodiversity," which takes a humanistic, improvisational, neurodiversity paradigm-informed approach to music therapy with two young autistic individuals with profound learning disabilities [13]. Pickard repeatedly emphasises that, in line with the neurodiversity paradigm, she is aiming to "enable and empower growth…from a non-normative perspective" (p. 303), "nurture authentic communication strategies" (p. 311) and provide "opportunities for growth to emerge through a non-directive therapeutic relationship" (p. 312). As she points out, the experience of autistic people with profound disabilities has been neglected in music therapy research. By illustrating the application of the neurodiversity paradigm to this client group, she demonstrates a deep sense of respect for the neurodivergent methods of interaction and communication she encounters [13].

Leza (an autistic music therapist) writes about "neuroqueering" music therapy, advocating that music therapists "approach autistic and neurodivergent people with a sense of cultural humility, and honour disabled people as experts in their own needs and experience" ([10] p. 218). She reminds readers of music therapy's roots as a deeply radical discipline with the potential to respond creatively to new ways of thinking, such as the neurodiversity paradigm. Leza gives a series of suggestions for neurodiversity-informed practice, including the importance of educating oneself about autistic culture, reorienting goals away from normalisation and positively responding to stimming and other autistic behaviours (such as in her descriptions of "Egg-pouring Song" and "Lining-Up-Music-Art-Song") [10].

Davies explores autism, music therapy and social justice in the context of the current trajectory of growing engagement with issues of equality, diversity, inclusion and belonging in UK music therapy discourse [11]. She suggests links between neurodiversity paradigm-informed music therapy and anti-oppressive music therapy [8], looking at power dynamics within the therapeutic relationship (particularly in cross-neurotype situations [37] – e.g., an allistic music therapist with an autistic client), the redefinition of autism as culture rather than pathology, and the acceptance of autistic characteristics such as stimming and echolalia as natural forms of expression that can be incorporated into musical improvisations. She looks ahead to the possibility of "neurodiversity-affirmative music therapy" (an approach also advocated for by Low et al. [38]), suggesting setting goals for therapy that seek to enhance "strengths, confidence, self-acceptance and coping strategies" [11]. In addition, she highlights the importance of considering minority stress [9], the double empathy problem [37] and the impact of trauma and co-occurring mental health difficulties when working with autistic people in music therapy.

In the past several years, the concepts of neurodiversity and the neurodiversity paradigm have received increasing attention within the music therapy literature. However, these concepts are not always well-developed within research: at times, a description of pathology paradigm music therapy practice may be preceded or succeeded by references to neurodiversity paradigm thinking, with little evidence of a connection being made between the two [39, 40]. In other examples of music therapy literature, particularly those that bear the influence of autistic people themselves (either through including an autistic co-author, or consulting autistic participants or family members of autistic people), the concepts of the neurodiversity paradigm can be seen to be more integrated with music therapy practice [41, 42].

The 2022 Cochrane Review of "Music Therapy for Autistic People" describes a "social or cultural model" approach to music therapy work, and it is clear from the description that this approach is consistent with the neurodiversity paradigm [43]. The neurodiversity paradigm is not directly mentioned in the review, although related concepts such as autistic characteristics being part of human diversity, the double empathy problem, neurotypical ways of socialising and autistic masking are alluded to. Ultimately, the review stays mainly within a pathology paradigm approach, exemplified by the use of pathology paradigm language such as "autism symptom severity" and the measurement of largely behavioural outcomes such as social interaction [43]. The fact that neurodiversity paradigm ideas are included and discussed, however, implies an increasing prominence of such approaches in contemporary music therapy discourse.

Ethnographic and Related Approaches

Research and applied work navigating the music-autism-neurodiversity nexus in music fields outside of music therapy – most notably in ethnomusicology, other sectors of the musicology discipline, and music theory – has at once formed a distinct body of literature and one that is in many ways inextricable from its music therapy counterparts. Inspired by the field of disability studies, the music theorist Joseph Straus [44] and composer Alex Lubet [45] wrote ground-breaking books that investigated diverse disability-related topics, including autism and neurodiversity. Edited volumes such as *Sounding Off: Theorizing Disability in Music* [46] and *The Oxford Handbook of Music and Disability Studies* [47] also helped to establish a foundational literature in this arena.

The first full-scale applied music project coming from the ethnomusicological side of the equation was the Music-Play Project, which ran from 2005 to 2009, was spearheaded by ethnomusicologists Michael Bakan and Benjamin Koen of Florida State University (FSU) and involved an interdisciplinary team of FSU collaborators including autism researchers, medical doctors, and psychologists [48]. The later Artism Ensemble and Speaking for Ourselves projects coordinated by Bakan represented subsequent phases of this work [16, 49, 50]. Prior to and subsequent to these initiatives, scholars and advocates including Shore [51], Dell'Antonio and Grace [52], Fessenden [53], Msumba [54] Leroy Moore and Keith Jones (Krip Hop Nation), and Elizabeth McLain have made essential contributions.

Case Examples

Case 1: Music Therapy (Hilary Davies)

P came to music therapy looking for support around difficult relationships (both past and present) with some of their family members. P had recently discovered they were autistic. They were in their forties and had a supportive partner and a professional job. P felt that they would like to experience a wider range of emotions, including more excitement, and also wanted to explore how music therapy could help them with processing difficult emotions related to both past and present experiences. P had played classical piano earlier in their life.

P was initially quite cautious and self-critical when we improvised together, but over time, with my reassurance that it did not matter what their music sounded like, P became more confident when improvising, which enabled their music to become more varied, with a greater range of pitches, rhythms and articulations a wider emotional range as well. This increased sense of comfort with "being themselves" within the music, along with a growing ability to open up to a wider range of emotions, reflected P's increasing self-confidence and emerging acceptance of all aspects of their autistic identity.

Over time, as P became more accepting of their autistic identity, including both the positive aspects and the challenges, they were also able to explore becoming more assertive about their needs and strengths in both professional and personal relationships, and additionally to begin to reframe some of their past and present experiences through an understanding of their autistic identity.

Case 2: Ethnomusicology (Michael Bakan)

Eight-year-old "Zena" thrived in the Music-Play Project (MPP). MPP involved free, improvisatory music-play sessions on a unique E-WoMP, or Exploratory World Music Playground. Each play group included three or four autistic children, co-participating parents, and ethnomusicologists Michael Bakan and Benjamin Koen as music-play facilitators [49]. The children played freely on a diverse range of E-WoMP percussion instruments, with no predefined musical outcomes or restrictions. Their parents were instructed to follow their lead without intervening (except for safety reasons). The facilitators used their expertise as music improvisers to nurture musical/social connections.

Zena was a very active E-WoMP player in MPP: energetic, socially engaged, and able to shift fluidly between leader and follower roles. A couple of years later, she and her parents enrolled in a new program coordinated by Michael: the Artism Ensemble project. Artism included six music-play facilitators rather than just two. All were accomplished musicians/ethnomusicologists from diverse nations (Trinidad, China, Bolivia, Peru, etc.). In contrast to the play-lab environment of MPP, Artism was a public-facing enterprise, performing concerts and workshops.

Artism's first rehearsal took place in January of 2011. Zena's mode of participation – or lack thereof – greatly surprised Michael. She spent almost the entire session staring blankly at a wall, stimming by flapping her hands, repeatedly bending and straightening her legs, pulling on her fingers, and occasionally standing up to pace stiff-legged around the room. She barely played a note and refrained from any significant interaction with the other players.

The second rehearsal followed suit. Afterwards, Zena's mother, Suzanne, took Michael aside. She explained that while Zena was enjoying Artism, the larger group size was overwhelming her: she needed time and space to take everything in and process it her own way, which for now meant mainly stimming. She was not ready to play the instruments, might never be, and did not want to be coaxed into doing so. Michael communicated all of this to the group, who willingly complied.

The third rehearsal witnessed a sea change in Zena's mode of engagement. Her stimming seemed more relaxed, she was less self-conscious, and she voluntarily played instruments and interacted with other group members. Her level of participation – both musical and social – progressively increased with each subsequent session, though she maintained a bimodal approach, with alternating episodes of what

Michael came to refer to as "playing mode" vs. "stimming mode," throughout the programme.

In May of 2011, about 6 weeks after the season's final Artism concert, Michael met with Zena and Suzanne at their home and asked Zena why, unlike in the Music-Play Project, she had been so reluctant to play instruments in Artism initially.

"Well, there was a reason why I played a lot in [MPP]," Zena responded. "I was afraid that someone was going to tell me I had to play if I didn't…. There were people with video cameras. It was just a lot of pressure and I felt like I had to play the instruments, like [that] was why I was there."

As for her initial resistance to actively playing in Artism two years later, Zena provided the following explanation: "[In] the Artism project, I have characters in my head. I think about them a *ton*, like probably more than I think about my own life. That's fine with me because they kind of relate to me. A lot of them have similar diagnosises *[sic]*…. Yeah, there's this one band of brothers…. There are ten of them…. And what's happening was, they were all musicians, the people in my head, and so I was imagining them playing the instruments, like I had one on the [Chinese] *zheng* [zither] and one on the *djembe* [drum], and everything…. At the beginning I was a little nervous that I'd have to play like in the last one [i.e., MPP 2009]. But after a while I realized it was cool if I could just express myself in any way. And in the end I felt comfortable enough and my characters kind of merged with it. That's when I started playing more."

"… So when you say the characters merged," Michael asked Zena, "did they become" – he paused – "you?"

Zena happily confirmed that this was the case. She seemed to have gained reassurance, particularly from Michael's responses to both her participation in the group and her verbal explanations of that participation, that having those characters in her head was just fine, that it was okay for them to merge and become her, and that it was okay for her to just be herself – on her own terms and not anybody else's.

And it was.

Conclusion and Future Considerations

In this chapter, we have examined how the emerging concept of the neurodiversity paradigm has both necessitated and inspired new ways of thinking, doing and acting, and has in turn significantly impacted theory and practice in both music therapy and ethnomusicology. Despite the different opportunities, challenges, and responsibilities afforded to music therapists and ethnomusicologists, a neurodiversity paradigm-informed approach involves a mutual and abiding commitment to the same core values and priorities across disciplines. These include the notion that all people, including autistic people, are the world's leading experts at being who they are, and that reciprocity and learning from – rather than intervention and teaching to – are of fundamental importance, regardless of the specific field of work.

We offer our readers both a model of collaborative possibility and an invitation to expand the horizons of such possibility. In doing so, we hope to inspire others in the music disciplines and beyond to do likewise.

This work has only just begun, yet it has already taken us quite far. To think and be in music and mind through the kaleidoscopic lens of neurodiversity is to be part of an exciting, promising, and yet largely uncharted future. We invite you to join us in the journey ahead.

References

1. Singer J. Neurodiversity: the birth of an idea. Kindle Edition; 1998.
2. Walker N. Throw away the master's tools: liberating ourselves from the pathology paradigm. In: Bascom J, editor. Loud hands: autistic people, speaking. Washington, DC: The Autistic Press; 2012.
3. Walker N. Neuroqueer heresies: notes on the neurodiversity paradigm, autistic empowerment, and Postnormal possibilities. Fort Worth: Autonomous Press; 2021.
4. Devlin K. How do I see you, and what does that mean for us? An autoethnographic study. Music Ther Perspect. 2018;36(2)
5. Cameron C. Does disability studies have anything to say to music therapy? And would music therapy listen if it did? Voices World Forum Music Ther. 2014;14(3)
6. Ansdell G. Community music therapy & the winds of change. Voices World Forum Music Ther. 2002;2(2)
7. Hadley S. Dominant narratives: complicity and the need for vigilance in the creative arts therapies. Arts Psychother. 2013;40(4):373–81.
8. Baines S. Music therapy as an anti-oppressive practice. Arts Psychother. 2013;40(1):1–5.
9. Botha M, Frost DM. Extending the minority stress model to understand mental health problems experienced by the autistic population. Soc Ment Health. 2020;10(1):20–34.
10. Leza J. Neuroqueering music therapy: observations on the current state of neurodiversity in music therapy practice. In: Milton D, editor. The neurodiversity reader. New York: Palgrave Macmillan; 2020. p. 210–25.
11. Davies H. "Autism is a way of being": an insider perspective on neurodiversity, music therapy and social justice. Br J Music Ther. 2022;36(1)
12. Pickard B, Thompson G, Metell M, Roginsky E, Elefant C. It's not what's done, but why it's done. Voices World Forum Music Ther. 2020;20(3)
13. Pickard B. Valuing neurodiversity: a humanistic, non-normative model of music therapy exploring Rogers' person-centred approach with young adults with autism spectrum conditions. In: Coombes E, Dunn H, Maclean E, Mottram H, Nugent J, editors. Music therapy and autism across the lifespan: a spectrum of approaches. London: Jessica Kingsley Publishers; 2019. p. 297–330.
14. Shiloh CJ, Blythe Lagasse A. Sensory friendly concerts: a community music therapy initiative to promote neurodiversity. Int J Community Music. 2014;7(1)
15. Bakan M. Ethnomusicological perspectives on autism, neurodiversity and music therapy. Voices World Forum Music Ther. 2014;14(3)
16. Bakan M. The musicality of stimming: promoting neurodiversity in the ethnomusicology of autism. MUSICultures. 2014;41(2):133–61.
17. Carrico AH. Constructing a two-way street: an argument for interdisciplinary collaboration through an ethnomusicological examination of music therapy, medical ethnomusicology, and Williams syndrome. Voices World Forum Music Ther. 2015;15(3)

18. Edwards J, Melchor-Barz GF, Binson B. Collaborating together: finding the emergent and disruptive in and between the fields of music therapy and medical ethnomusicology. Voices World Forum Music Ther. 2015;15(3)
19. Health and Care Professions Council. The standards of proficiency for arts therapists [Internet]. 2013 [cited 2023 Jan 18]. Available from: https://www.hcpc-uk.org/standards/standards-of-proficiency/arts-therapists
20. APA. DSM-5 desk reference. In: Diagnostic and statistical manual of mental disorders. Arlington: Author; 2013.
21. Oliver M. Understanding disability. London: Red Globe Press; 1996.
22. Barrington KA. Music therapy: a study in professionalisation. Durham University; 2005.
23. Sinclair J. Don't mourn for us. Our Voice. 1993;1(3):1–7.
24. Alvin J. Music therapy for the autistic child. Oxford: Oxford University Press; 1978.
25. Nordoff P, Robbins C. Therapy in music for handicapped children. London: Gollancz; 1971.
26. Robarts J. Music therapy for children with autism. In: Trevarthen C, Aitken K, Papoudi D, editors. Children with autism: diagnosis and interventions to meet their needs. London: Jessica Kingsley Publishers; 1998. p. 134–60.
27. Martin L. Applied behaviour analysis: introduction and practical application in music therapy for young children with autism spectrum disorders. In: Kern P, Humpal M, editors. Early childhood music therapy and autism Spectrum disorders: developing potential in Young children and their families. Philadelphia: Jessica Kingsley Publishers; 2012. p. 101–16.
28. Dunn H, Coombes E, Maclean E, Mottram H, Nugent J, editors. Music therapy and autism across the lifespan: a spectrum of approaches. London: Jessica Kingsley Publishers; 2019.
29. Kern P, Rivera N, Chandler A, Humpal M. Music therapy Services for Individuals with autism spectrum disorder: a survey of clinical practices and training needs. J Music Ther. 2013;50(4):274–303.
30. Wigram T. Indications in music therapy: evidence from assessment that can identify the expectations of music therapy as a treatment for autistic Spectrum disorder: meeting the challenge of evidence based practice. Br J Music Ther. 2002;16(1):11–28.
31. Bascom J. Quiet hands. In: Bascom J, editor. Quiet hands: autistic people speaking. Washington, DC: Autism Self-Advocacy Network; 2012. p. 177–82.
32. Chance P. After you hit a child, you can't just get up and leave him; you are hooked to that kid. O. Ivar Lovaas interview with Paul Chance; 1974. http://neurodiversity.com/library_chance_1974.html
33. Bieleninik Ł, Geretsegger M, Mössler K, Assmus J, Thompson G, Gattino G, et al. Effects of improvisational music therapy vs enhanced standard care on symptom severity among children with autism spectrum disorder: the TIME-A randomized clinical trial. JAMA – J Am Med Assoc. 2017;318(6)
34. Stige B. The relentless roots of community music therapy. Voices World Forum Music Ther. 2002;2(3)
35. Young L. Finding our voices, singing our truths: examining how quality of life domains manifested in a singing group for autistic adults. Voices World Forum Music Ther. 2020;20(2)
36. Neurotribes SS. The legacy of autism and how to think smarter about people who think differently. Crow's Nest: Allen & Unwin; 2015.
37. Milton DEM. On the ontological status of autism: the "double empathy problem". Disabil Soc. 2012;27(6):883–7.
38. Low MY, Devlin K, Sofield S. A response to Blauth and Oldfield's "Research into increasing resilience in children with autism through music therapy: statistical analysis of video data". Nord J Music Ther. 2022;31(5):481–3.
39. Balducci A. Forms of vitality and microanalysis in music therapy within adult autism: a clinical report. J Music Ther. 2021;13(2)
40. Vlachová Z. Means of musical dialogues and reciprocity. Voices World Forum Music Ther. 2022;22(2)

41. Low MY, McFerran KS, Viega M, Carroll-Scott A, McGhee Hassrick E, Bradt J. Exploring the lived experiences of young autistic adults in Nordoff-Robbins music therapy: an interpretative phenomenological analysis. Nord J Music Ther. 2022;32(4):341–64.
42. Mössler K, Halstead J, Metell M, Gottschewski K, Schmid W. The room is a mess: exploring the co-creation of space for attunement dynamics between an autistic child and a non-autistic music therapist. Nord J Music Ther. 2022;32(4)
43. Geretsegger M, Fusar-Poli L, Elefant C, Mössler KA, Vitale G, Gold C. Music therapy for autistic people, Cochrane database of systematic reviews, vol. 2022. John Wiley and Sons Ltd; 2022.
44. Straus J. Extraordinary measures: disability in music. New York: Oxford University Press; 2011.
45. Lubet A. Music, disability, and society. Philadelphia: Temple University Press; 2011.
46. Lerner N, Straus JN. Sounding off: theorizing disability in music. New York: Routledge; 2006.
47. Blake H, Jenson-Moulton S, Lerner N, Strauss JN, editors. The Oxford handbook of music and disability studies. New York: Oxford University Press; 2016.
48. Bakan MB, Koen B, Kobylarz F, Morgan L, Goff R, Kahn S, et al. Following frank: response-ability and the co-creation of culture in a medical ethnomusicology program for children on the autism spectrum [Internet]. Vol. 52, Source: Ethnomusicology, Spring/Summer. 2008. Available from: https://www.jstor.org/stable/20174586
49. Bakan, M. 'Don't go changing to try and please me: combating essentialism through ethnography in the ethnomusicology of autism. Ethnomusicology. 2015;59(1):116–44.
50. Bakan M, Chasar M, Gibson G, Grace EJ, Hamelson Z, Nitzberg Z, et al. Music & Autism: speaking for ourselves. New York: Oxford University Press; 2021.
51. Shore S. Beyond the wall: personal experiences with autism and Asperger syndrome. 2nd ed. Shawnee Mission: Autism Asperger Publishing Company; 2003.
52. Dell'Antonio A, Grace EJ. No musicking about us without us! J Am Musicol Soc. 2016;69(2):553–9.
53. Fessenden JW. Autistic musicality: history, theory, and two case studies. Stony Brook: State University of New York; 2020.
54. Msumba J. Shouting at leaves. BookBaby/Jennifer Msumba; 2021.

Psychosocial Aspects of Music Therapy

16

Amanda Rosado and Rebecca Vaudreuil

Introduction

Constructs of human identity, expression, and connection along with the intersectionality of music and neuroscience research underpin the impact of music and the profession of music therapy (MT). MT is the clinical use of music experiences by a credentialed professional who has completed an approved music therapy program [1]. Music experiences can provide abundant and underutilized resources for health and well-being to diverse age and cultural groups, and those with varying backgrounds. This is accomplished by informing a broad array of music practices and experiences that may elicit biopsychosocial and spiritual effects. If exposure to and engagement in music can change brain patterns, shape behavior, and enhance social connectedness, it can certainly support psychosocial functioning and development on a systems level from the individual/self (i.e., micro) to families and communities (i.e., mezzo) to societal and global contexts (i.e., macro). For example, music therapists help clients with emotional regulation and expression, social connection, identity, and biopsychosocial needs [2–8]. Recent treatment findings specific to

Disclaimer: The opinions contained herein represent the private views of the authors and are not to be construed as official or as reflecting the views, opinions, or policies of the Henry M. Jackson Foundation for the Advancement of Military Medicine, Inc., the National Endowment for the Arts, Department of Defense or the U.S. Government. Mention of trade names, commercial products, or organizations does not imply endorsement by the US Government. This material was created free of branding or market affiliations. The authors are operating solely as contributors.

A. Rosado (✉)
Independent Author, Germantown, MD, USA

R. Vaudreuil
Henry M. Jackson Foundation for the Advancement of Military Medicine, Inc. in support of Creative Forces®: NEA Military Healing Arts Network, Bethesda, MD, USA
e-mail: rebecca_vaudreuil@mail.harvard.edu

© The Author(s), under exclusive license to Springer Nature Switzerland AG 2023
K. Devlin et al. (eds.), *Music Therapy and Music-Based Interventions in Neurology*, Current Clinical Neurology,
https://doi.org/10.1007/978-3-031-47092-9_16

individuals with neurological disorders emphasize the importance of addressing psychosocial experiences as part of the holistic care continuum [9, 10]. Music therapists working with neurological disorders identify that music accessibility and familiarity can improve quality of life. Studies have shown this through reporting improvements in pre- and post- mental and physical measures that include social functioning [11].

This chapter presents two music therapists' experiences working in neurologic rehabilitation and facilitating psychosocial support via music experiences in individual and group contexts, both in-person and remotely. The authors discuss psychosocial needs in person- affirming spaces through literature review, defining and distinguishing components of psychosocial support, providing case examples and protocols, and outlining future directions for inclusive practice.

Self-Locations

In her music therapy practice, Amanda has supported folks with mental health needs, neurological disorders, neurodivergence, and end of life support. In her practice, she finds herself gravitating toward addressing the 'bigger picture' in therapy spaces, collectively processing the impact of the diagnosis, associated grief, and celebration of resilience. Amanda identifies as white, Latinx, queer, able-bodied, middle-class, and neurotypical. She is a long-time musician, which is a part of her identity and her wellness. She acknowledges that the field of music, including MT, is heavily rooted in Eurocentric culture. Through acknowledging gate-keeping, ableism, cultural appropriation, and white supremacy in the field of MT, she upholds an intention to maintain spaces that encompass emotional and physical safety, and accessibility. She commits to learning and unlearning, welcoming clinical spaces that empower collaboration, curiosity, transparency, and honesty in therapeutic relationships.

Rebecca has facilitated music therapy in medical, mental health, military/veteran, educational, neurologic rehabilitation, and community settings. After years of music therapy clinical practice, she pursued advanced degrees in educational neuroscience and social work, respectively, to better understand the intersection of music and health across diverse individuals, communities, and environments. In 2014, Rebecca joined an initiative of the National Endowment for the Arts (NEA), Creative Forces®: NEA Military Healing Arts Network, which is managed in partnership with Americans for the Arts, Civic Arts, the Henry M. Jackson Foundation for the Advancement of Military Medicine, and Mid-America Arts Alliance. She currently serves as the Creative Forces Lead Music Therapist and Clinician Supervisor wherein she provides clinical mentorship, supports increased access to creative arts therapies research across military and veteran healthcare systems, and promotes collaborative arts practices in clinical and community settings. Ultimately, she aims to create integrative approaches that leverage the arts to enhance healthcare in a way that respects, encourages, and celebrates individual, community, societal, and global diversity.

These authors are dedicated to using language aligned with and reflective of anti-ableist practices, as they take a strengths-based rather than a prescriptive approach to treatment. Therefore, MT 'interventions' are referred to less prescriptively as MT 'experiences', and 'disorders' and 'diagnoses' are not used to identify individuals or communities impacted by them.

Definitions and Background

Psychosocial Definitions in Healthcare

To comprehensively discuss psychosocial aspects of MT experiences in neurology, a first step is to unpack its meaning across healthcare professions. In the medical field, "psychosocial" is defined as "the mental, emotional, social, and spiritual effects of a disease…changes in how a patient thinks, their feelings, moods, beliefs, ways of coping, and relationships with family, friends, and coworkers" ([12], para 1). From the psychological perspective, the American Psychological Association defines psychosocial as the "intersection and interaction of social, cultural, and environmental influences on the mind and behavior" [13]. The National Association of Social Work operates on a bio-psycho-social-spiritual framework and "recognizes the importance of whole person care and takes into account…physical or medical condition; emotional or psychological state; socioeconomic, sociocultural, and sociopolitical status; and spiritual needs" ([14], p. 10). Furthermore, the International Federation Reference Centre for Psychosocial Support defines psychosocial as a "process of facilitating resilience within individuals, families, and communities by respecting their independence, dignity and coping mechanisms" ([15], p. 25).

Neurological disorders can create a rift in personhood, identity, functioning, and quality of life. The adoption, adaptation, and integration of psychosocial approaches across healthcare fields acknowledges the usefulness of person-centered and strengths-based practices.

Psychosocial Support Operationally Defined

It is difficult, if not impossible to separate bio, psycho, and social domains when processing lived experiences, especially pertaining to treatment approaches that employ creative processes. The authors of this chapter created an operational definition of "psychosocial support" informed through their clinical work and review of diverse classifications (i.e., medicine, psychology, social work), as demonstrated in Fig. 16.1. Our definition aims to contextualize and highlight the use of music experiences to address psychosocial needs across treatment areas, functional and behavioral domains, and life stages of individuals with neurological disorders.

We define psychosocial support as the development, continued growth, and interaction of individuals and communities across systems levels (i.e., micro, mezzo,

Fig. 16.1 Common themes from cross-discipline psychosocial literature, as visualized by the authors using freewordcloudgenerator.com

macro) through creative processes that promote inclusion, connection, collaboration, and cohesion while acknowledging diversity as a strength. Other terms to be mindful of throughout the chapter are acceptance, identity, self-worth, resilience, integration, grief, and vulnerability.

Cue the Music: Music and Psychosocial Considerations

Culture, environment, and social influences impact psychosocial development, behavior, and functioning and music is a key player across these areas; therefore, it is not surprising that music can inform, influence, and make connections between them. MT experiences can be categorized into different methods including receptive (e.g., active music listening), recreative (e.g., playing/singing pre-composed music), and creative (e.g., improvisation, songwriting) [16]. These methods, including their perceived influences on psychosocial development, will be further examined in this chapter through review of the literature and case examples.

Music in psychosocial training and counseling (MPC®) is a neurologic music therapy (NMT)® technique designed to utilize music-based methods to help people with neurologic issues improve their psychosocial functioning. MPC® uses "musical performance to address issues of mood control, affective expression, cognitive coherence, reality orientation, and appropriate social interaction" ([17], p. 197). The primary goals of MPC® can be loosely divided into social competence, mood control, self-awareness, and affect identification and expression [18].

Literature Review

Neuroscientific research on the expressive arts, specifically music and dance, is becoming increasingly popular due to the perceived positive impacts on health and well-being. A scoping review by Dingle et al. [19] indicated that arts engagement can be linked to improvements across multiple health domains. For example, they found that shared music listening and group singing enhanced social connectedness and mood for hospitalized older adults. Group singing supported cognitive health, and active music making on instruments had similar impacts in social and cognitive domains including improved motor functions in school students, older adults, and people with mild traumatic brain injury (See Chap. 4).

Social and cultural inclusion, connection, self-esteem, and empowerment through rapping, songwriting, and music composition may increase well-being in marginalized communities. In some studies, receptive and intentional music listening reduced pain perception through changes in physiological arousal. Movement to music is associated with improved health and well-being in people who live with dementia, postnatal depression, and obesity through various cognitive, physical, and social processes [19].

Music Therapy to Address Psychosocial Needs

Emerging research highlights the importance of meeting psychosocial needs of individuals with neurological disorders, especially considering the prevalence of comorbid mood disorders and dysregulation that often accompany neurological disorders and are historically undertreated [10, 20, 21]. Neurological disorders may also lead to difficulty in identifying, expressing, and processing emotions including grief and loss of ability. Many music therapists design music experiences to address client goals, which often helps meet needs across treatment domains [3, 5]. MT participants report increased awareness, regulation, and expression of emotions through the use of music in a therapeutic space [2–7]. Living with neurological disorders can be an isolating experience for individuals and their families; engagement in MT can combat these feelings and help build a sense of belonging among peers [5, 7, 8]. Additionally, studies suggest that MT promotes identity building and reintegration across familial, occupational, and social domains [3, 5, 8], as well as increases in positive connections with caregivers [5, 7].

Music Therapy Experiences for Psychosocial Goals

From the myriad of MT experiences implemented in neurological settings, clinicians report that songwriting, music making/improvisation, receptive experiences, and therapeutic performance are most helpful in meeting the psychosocial goals of clients [2, 3, 22–24]. Many clinicians note that these experiences can be particularly impactful in group settings, by building a sense of community and belonging.

Songwriting in therapeutic spaces creates a context in which the client can take ownership and gain a deeper understanding of their experiences [25]. For the neurological community, songwriting has been meaningful in identifying themes related to loss, acclimating to new norms, and processing injury and/or incident-related traumas [3, 5, 8]. Additionally, songwriting- individually and in group contexts- may help clients reconnect to their sense of self by facilitating reminiscence, emotional connection, and belonging [3].

Many music therapists employ live music making in group settings to support psychosocial goals [3, 5, 8, 24]. Clinicians identify that mood regulation occurs through the process of rhythmic entrainment – humans' natural physical and psychological alignment with rhythm [26]. Entrainment can promote emotional catharsis, nervous system regulation, stress reduction, and emotional regulation [3, 24]. In collaborative music making with clients, the iso-principle is often used in improvisational spaces, in which music therapists match music to clients' presenting mood states and then modulate the music intentionally to shift the emotional space [27]. The music therapist is attuned to the client throughout the facilitation of the iso-principle process, responds accordingly, and can encourage a sense of understanding and empathy that many individuals find beneficial in mental health recovery [28]. Shared understanding of these processes is further emphasized through group music making, which encourages community-building and shapes a supportive and emotionally safe social structure [3, 5, 22, 24].

Receptive music experiences, such as music-assisted relaxation, music and imagery, and active music listening, reportedly help meet emotional needs of clients in group and individual contexts [3, 5, 24]. Sharing and discussion of preferred music can support clients to use music listening in their daily lives as a self-care strategy and a healthy coping tool [3]. Music-assisted relaxation, in which the therapist provides live music while facilitating breathwork, progressive muscle relaxation, grounding, or imagery exercises, supports in-the-moment physiological and emotional regulation.

Many music therapists identify therapeutic music performance as helpful in supporting clients in establishing a sense of accomplishment and connection with themselves and others [29, 30]. Audiences are often composed of family, friends, fellow clients, providers, and community members. Therapeutic performance fosters a unique element of connection by viewing people through a strengths-based lens as creative, healthy, and empowered. This performance-based model exists on a clinic-to-community continuum that supports shifts from individual to group MT sessions ultimately moving towards independent arts engagement. Clients who participate in performance often report feeling supported in acknowledging their own strengths, gaining a deeper understanding of the benefits of MT, and heightening awareness of how to integrate into their communities [22].

Case Vignettes

Case Vignette #1: John's Journey (Contributed by Amanda)

John was one of my first private practice clients, and I worked with him for about 2 years. John was 69 years old, white, retired from government work, a husband, and a father of two adult children. With a diagnosis of Parkinson Disease (PD), John started MT after his wife learned about the benefits of music experiences on managing Parkinsonian symptoms. John was a long-time blues harmonica (i.e., harp) player, band member, and singer-songwriter. Textbox 16.1 contains lyrics from one of John's original songs, "Shotgun Wedding."

> **Textbox 16.1: "Shotgun Wedding"** *(an excerpt)*
> I went on down to Louisiana/It was the first of May
> Looking for some red beans and rice and a crawfish etouffee
> I met a woman in the port-a-call, she said I was the one
> We went back to her house, and man we had some fun
>
> In walked her brother and her daddy too, coon guns in their hands.
> They said, "Welcome to the family son, or else you'll be a dead man"
>
> It was a shotgun wedding in a shotgun house, down in the voucaree
> Me and Evangelina, we met on our wedding day

He started MT with a self-reported goal of "vocal restoration," recognizing how his voice was impacted by PD. He came to sessions with a large binder of original songs and a large collection of harps in all keys. With John's goal in mind, our sessions initially focused on vocal warmups and singing. He expressed that some days were more successful than others. He also often expressed quiet frustration when he confused specific words in his songs, though there were many times when he adapted and changed words in the moment. Most of his songs were upbeat, humorous, and full of imagery.

When things got frustrating, we shifted to blues improvisation, with John playing harp while I accompanied on guitar. Things shifted as his frustrations grew, and this was exacerbated in 2020 due to the mandated lockdowns from the COVID-19 pandemic. Although John liked to try to stick to the plan of singing and music making together in virtual sessions, he shared many moments where he was reckoning with his loss of abilities.

At this point, session intentions changed based on John's emotional needs. A major tone shift was when he shared one of his songs, "I'm So Sorry." The style of this was starkly different compared to the rest of his upbeat, humorous songs. He originally sang the song a cappella, and a MT intern notated accordingly. Textbox 16.2 displays lyrics from "I'm So Sorry".

> **Textbox 16.2: "I'm So Sorry"** *(an excerpt)*
> I'm so sorry baby, I made a big mistake
> I thought I needed her, but it only caused heartache
> I went out on you darling, and I gave my love away
> Now I'm asking for your forgiveness, please hear what I have to say
>
> I'm so sorry I'm so sorry
> Believe me when I say, I wanna come right home to stay

John processed these lyrics in a session, reflecting on past mistakes. John's focus moving forward involved reflection and reminiscence of life experiences, which centered around visiting New Orleans, teaching children harp, and sharing memories of his band. He often gravitated toward harp improvisation instead of focusing on his songs, even with the audio latency that accompanies virtual sessions.

Near the end of our time working together, John's pancreatic cancer, which was in remission up to this point, returned with an aggressive and terminal diagnosis. As he moved to in-home hospice care, surrounded by friends and family, sessions leaned toward reflection, reminiscence, and meaningful connections. An intern and I re-created his written songs, as he was unable to, as a way for the family to connect to him through his identity as a blues musician.

Session spaces shifted throughout our time together. The MT experiences moved from skills-oriented goals to reckoning with loss, to celebrating John's personhood. There was an organic transition to an inevitable focus on holding space for the emotions that his music experiences brought up for him and his loved ones.

Case Vignette #2: Psychosocial Music Skills Group Protocol for U.S. Veterans Experiencing Complex and Comorbid Neurological and Psychological Health Conditions (Contributed by Rebecca)

My current clinical MT work includes facilitating individual and group sessions for U.S. Veterans conducted exclusively through telehealth. This work is supported by Creative Forces and aligned with their mission to improve the health, well-being, and quality of life for military and veteran populations exposed to trauma. Creative Forces places creative arts therapies at clinical sites throughout the country, including telehealth services, and increases access to community arts engagement [31].

This case vignette presents a virtual Music Skills group that provides creative space for rural veterans who experience comorbid neurological (e.g., traumatic brain injury), psychological (e.g., post-traumatic stress, anxiety, depression), and other health conditions. Participants are referred to this group upon or near completion of individual MT treatment as part of the clinical pathway, shown in Fig. 16.2.

16 Psychosocial Aspects of Music Therapy

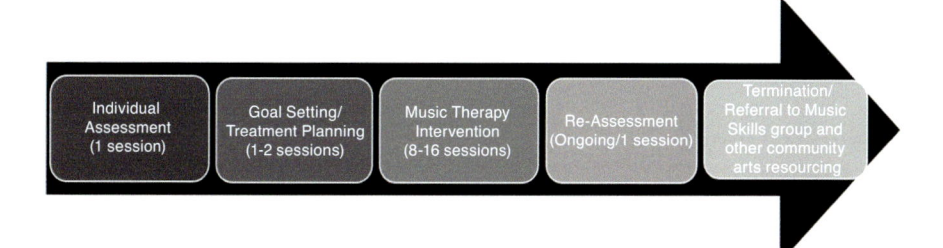

Fig. 16.2 Music therapy clinical pathway

Table 16.1 Music Skills Group Structure (60 and 120 min)

Group goals	(i) Promote appropriate socialization (ii) Encourage peer support and engagement (iii) Decrease stress and anxiety (iv) Increase coping skills and strategies for reintegration (v) Support memory/recall
Group contracting	(i) Keep camera on (ii) Engage consistently, as able (iii) Take turns speaking (iv) Remain muted when practicing and/or if the environment is noisy (v) Come to session prepared (e.g., independent skills practice) (vi) Indicate an emergency or change in environment via pre-determined gesture/process.
Group structure: 60-min group	(i) Check-in/introductions (10 min) (ii) Music skills review and resource sharing (10 min) (iii) Music therapist or participant (preferred) facilitated warm-up (5 min) (iv) Open creative engagement time/independent skills practice (25 min) (v) Check-out discussion/planning for next group (10 min)
Group structure: 120-min group	(i) Check-in #1 with introductions (10 min) (ii) Music skills review and resource sharing (10 min) (iii) Music therapist or participant (preferred) facilitated warm-up (5 min) (iv) Independent skills practice #1: Open creative engagement time (15 min) (v) Check-in #2 (10 min) (vi) Independent skills practice #2: Open creative engagement time (20 min) (vii) Check in #3 (10 min) (vii) Independent skills practice #3: Open creative engagement time (30 min) (ix) Check-out discussion/planning for next group (10 min)

Music Skills group supports independent and group social skill-building through creative engagement, participatory action, psychoeducation, and can be adapted for use with various clientele. Isolation often accompanies complex and comorbid traumas; hence, this group is aimed at mobilizing participants from clinical MT to community-based arts engagement through didactic and experiential learning, information and resource sharing, and peer-support. This vignette provides a group description (e.g., goals, expectations, structure), replicable/adaptable protocols (e.g., 60 and 120-min), examples of group member participation, and outcomes

(See Table 16.1). Names have been changed in accordance with the Health Insurance Portability and Accountability Act.

Group Description

Virtual group sessions occur weekly for up to 120 min and there are up to 10 Veterans in each group. Group goals are aimed at: (1) promoting socialization; (2) encouraging peer support; (3) decreasing stress and anxiety; (4) supporting cognition; and (5) increasing coping skills and strategies for reintegration. Group contracting is co-constructed with participants, and includes keeping cameras on, engaging consistently, taking turns speaking, remaining muted when practicing and/or if their environment is noisy; being prepared, and indicating an emergency or change in environment via pre-determined gesture/process.

Group structure incudes (a) participant check-in/introductions, (b) music skills review and resourcing, (c) music therapist or participant (preferred) facilitated warm-up: (d) creative engagement time/independent skills practice, and (e) check-out/discussion/planning for next group, described below.

Participant Check-in and Introductions

Check-ins generate important feedback that informs the wellbeing of participants and initiates engagement of the group through sharing updates on musical accomplishments and continued needs. For example, participant "Sean" shared that he has continued to work on learning fundamental skills on the ukulele and learning a song. During check-in, Sean played a segment of the song he is learning for the group, which builds community and provides a baseline of Sean's progress that can be used to gauge improvement during sessions and from week-to-week.

Music Skills Review and Resourcing

Reviewing group structure and contracting ensures that participants understand session expectations and flow. The music therapist and participants share creative resources that are accessible in participants' communities as well as virtually. One participant, "Gary" shared his experience with an online program that provides guitar lessons for veterans and informed the group how to enroll in the class.

Warm-Up Exercise

The music therapist or participant (preferred) facilitates a warm-up that reinforces music awareness, creates connections, and promotes purposeful music selection and techniques for relaxation, motivation, focus, and other desired outcomes. A music therapist-led warm-up example is progressive muscle relaxation that uses musical dissonance to match muscle tension and uses musical consonance to work towards resolution and release. A participant may share a playlist created for specific purposes. For example, "Tammy" shared her motivational playlist and guided the group through her process of song selection. Songs on Tammy's playlist were upbeat in tempo, included various instrumentation, had positive messaging, and were both instrumental and lyrical. Certain songs reflected her life pre-injury, were reminiscent of her military service, and reminded her of how far she has come in treatment. Textbox 16.3 includes a sample of songs from Tammy's playlist.

Textbox 16.3: Motivational Songs

"Come and Get Your Love" by Redbone

"Lovely Day" by Bill Withers

"Another One Bites the Dust" by Queen

"Don't Worry, Be Happy" By Bobby McFerrin

"Respect" by Aretha Franklin

"Mercy" by Dave Matthews

"Claire de lune" by Claude Debussy

"Sunshine On My Shoulder" by John Denver

"Strength, Courage, and Wisdom" by India Arie

Creative Engagement Time/Independent Skills Practice

I often hear from clients that it is hard to prioritize time for creativity in their busy lives. This part of group holds space and time for participants to engage in independent creative practices (e.g., playing/writing songs, music learning, creative writing) and seek guidance and support as needed from the music therapist and peers. For example, "Sean" played the ukulele to reinforce motor memory and got help from peers to learn a riff, Gary" worked on guitar exercises and songs from his online guitar class, and "Tammy" re-wrote lyrics of a song from her playlist to express thoughts about a current life situation, which she shared with the group.

Check-out/Discussion and Planning

Participants share insights from group and plan for continued engagement between sessions, which includes setting practice benchmarks that support ongoing progress. Participants report that intentionality and accountability are essential in motivating creative practice between sessions and use of music-based strategies for biopsychosocial support.

Additional Psychosocial Supports

Office hours are offered outside of group time and provide an opportunity for consultation with a music therapist to address continued needs and facilitate a warm hand-off to community resourcing. A monthly resource sheet is disseminated to support continued creative engagement by compiling musical content such as apps, songs, mindfulness skills, practice techniques, and community and virtual programming shared during group.

Psychosocial Benefits

Participants report that Music Skills group provides a sense of community and motivates creativity. It holds space, time, and support to intentionally engage in creative processes that participants may not otherwise have, and collective group work is inspiring. Although group members have varying levels of musical experience, it is commonly reported that re-acclimating to music helps them reintegrate across different areas of their lives.

Conclusion and Future Directions

Regarding the future of treating neurological disorders, the authors posit the need for psychosocial approaches to be made a primary focus of treatment outcomes rather than an indirect benefit or secondary gain. This can be accomplished through education and training of music therapists and other healthcare professionals that position psychosocial skills as an essential component of rehabilitation. Additionally, it is important to build upon different MT approaches that support deeper and wider spaces in which clients can explore their emotions, identities, and sense of belonging within the experiences of neurological disability. Supporting this aspect of health is necessary when sharing space with someone who is navigating significant changes while also processing the impact of that change and how to adapt to it.

The literature review and case examples provided in this chapter reflect the inherent nature of music and its ability to meet the psychosocial needs of individuals and communities. In its truest essence, music facilitates social connection, self-attunement, and outward expression of emotion through alignment with the human condition. Humans are hardwired to biologically process and respond to sensory input such as sound, and softwired for auditory stimuli to elicit emotions and impact other psychosocial factors, which may further contextualize responses. Therefore, limiting the use of music to solely addressing concrete physiological and functional skills represents a grave disservice to the often-natural way that music can access deeper, necessary levels of processing lived experiences.

Another aspect to consider when thinking about the future of psychosocial approaches to MT is how concepts of 'acceptable functioning' have historically been labeled and categorized in society, which can be viewed as a product of ableist practices. Let us continue to ponder what it means to function with a disability and how and by whom functioning is defined; for example, by clients' self-appraisals and insights rather than on the basis of pre-conceived and often reductionist societal perceptions.

If we embrace the notion that psychosocial experiences are often under emphasized or overlooked when compared to neurologic rehabilitative goals (e.g., speech, motor, cognition), let this acknowledgement encourage the field of MT and healthcare-at-large to move towards actionable processes of identifying and addressing psychosocial challenges and needs. Furthermore, it is recommended to use client-centered and strengths-based models that encourage clients to trust their intuition and contribute to shaping the therapeutic space for their maximum benefit.

References

1. What is music therapy? [Internet]. American Music Therapy Association (AMTA); 2005. Available from https://www.musictherapy.org/about/musictherapy/.
2. Brancatisano O, Baird A, Thomson WF. Why is music therapeutic for neurological disorders? The therapeutic music capacities model. Neurosci Biobehav Rev. 2020;112:600–15.
3. Bronson H, Vaudreuil R, Bradt J. Music therapy treatment of active duty military: an overview of intensive outpatient and longitudinal care programs. Music Ther Perspect. 2018;36(2):224–33.
4. Chilton G, Vaudreuil R, Freeman EK, McLaughlan N, Herman J, Cozza SJ. Creative forces programming with military families: art therapy, dance/movement therapy, and music therapy brief vignettes. J Mil Veteran Fam Health. 2021;7(3):104–13.
5. Magee W. Why include music therapy in a neuro-rehabilitation team? Adv Clin Neurosc Rehabil. 2019; https://doi.org/10.47795/STUI1319.
6. Raglio A, Attardo L, Gontero G, Rollino S, Groppo E, Granieri E. Effects of music and music therapy on mood in neurological patients. World J Psychiatry. 2015;5(1):68–78.
7. Van Bruggen-Rufi M, Vink A, Achterberg W, Roos R. Improving quality of life in patients with Huntington's disease through music therapy: a qualitative explorative study using focus group discussions. Nord J Music Ther. 2018;27(1):44–66.
8. Vaudreuil R, Avila L, Bradt J, Pasquina P. Music therapy applied to complex blast injury in interdisciplinary care: a case report. Disabil Rehabil. 2019;41(19):2333–42.
9. Duits A, van der Heijden C, van Het Hoofd M, Roodbol G, Tiemessen M, Munneke M, Steppe M. Psychosocial needs of patients and spouses justify a position of psychosocial health professionals in the multidisciplinary care for Parkinson's disease. Clin Parkinsonism Relat Disord. 2020;3:1–4.
10. Pick S, Goldstein LH, Perez DL, Nicholson TR. Emotional processing in functional neurological disorder: a review, biopsychosocial model and review agenda. J Neurol Neurosurg Psychiatry. 2019;90(6):704–11.

11. McCray J, Altenmuller E, Scholz D. Association of music interventions with health-related quality of life: a systematic review and meta analysis. JAMA Netw Open. 2022;5(3):e223236. https://www.ncbi.nlm.nih.gov/pmc/articles/PMC8941357/#__ffn_sectitle
12. National Cancer Institute. Psychosocial: NCI dictionary of cancer terms. National Institute of Health. n.d. https://www.cancer.gov/publications/dictionaries/cancer-terms/def/psychosocial
13. American Psychological Association. Psychosocial. APA Dictionary of Psychology. 2022. https://dictionary.apa.org/psychosocial
14. Wheeler D, McClain A, Cox L, Fritz T, Virna L, Otis-Green S, Yamamoto A, Collins S. National Association of Social Work. NASW standards for social work practice in health care settings. 2016. https://www.socialworkers.org/LinkClick.aspx?fileticket=fFnsRHX-4HE%3D&portalid=0.
15. International Federation Reference Centre for Psychosocial Support. What We Do. 2021. Accessed from: https://pscentre.org/what-we-do/.
16. Bruscia KE. Defining music therapy. 3rd ed. University Park: Barcelona Publishers; 2014.
17. Thaut MH. Neurologic music therapy techniques and definitions. Rhythm, music and the brain: Scientific foundations and clinical applications. 2005.
18. Wheeler BL. Music in psychosocial training and counseling (MPC). In: Thaut MH, Hoemberg V, editors. Handbook of neurologic music therapy. Oxford University Press; 2014. p. 331–59.
19. Dingle G, Sharman L, Bauer Z, Beckman E, Broughton M, Bunzli E, Davidson R, Draper G, Fairley S, Farrell C, Flynn L, Gomersall S, Hong M, Larwood J, Lee C, Lee J, Nitschinsk L, Peluso N, Reedman S, Vidas D, Walter Z, Wright O. How do music activities affect health and well-being? A scoping review of studies examining psychosocial mechanisms. Front Psychol. 2021;12:713818. https://doi.org/10.3389/fpsyg.2021.713818.
20. Bono AD, Twaite JT, Krch D, Mccabe DL, Scorpio KA, Stafford RJ, Borod JC. Mood and emotional disorders associated with parkinsonism, Huntington disease, and other movement disorders. Handb Clin Neurol. 2021;183:175–96.
21. Yang S, Sajatovic M, Walter B. Psychosocial interventions for depression and anxiety in Parkinson's disease. J Geriatr Psychiatry Neurol. 2012;25(2):113–21.
22. Vaudreuil R, Bronson H, Bradt J. Bridging the clinic to community: music performance as social transformation for military service members. Front Psychol. 2019;10(119) https://doi.org/10.3389/fpsyg.2019.00119.
23. Vaudreuil R, Reuer B. *MusicWorx toolbox series:* an introduction to military music therapy programming: working with active service members and veterans. 3rd ed. San Diego: MusicWorx; 2013. isbn: 978-0-9909673-0-9
24. Vetro-Kalseth D, Vaudreuil R, Segall LE. Treatment description and case series report of a phased music therapy group to support veteran reintegration. Mil Psychol. 2021;33(6):446–52.
25. Baker F, Wigram T. Songwriting: methods, techniques and clinical applications for music therapy clinicians, educators and students. London: Jessica Kingsley Publishers; 2005.
26. Ross JM, Balasubramaniam R. Physical and neural entrainment to rhythm: human sensorimotor coordination across tasks and effector systems. Front Hum Neurosci. 2014;8:576. https://doi.org/10.3389/fnhum.2014.00576.
27. Starcke K, Mayr J, von Georgi R. Emotion modulation through music after sadness induction- the iso principle in a controlled experimental study. Int J Environ Res Public Health. 2021;18(23):12486. https://doi.org/10.3390/ijerph182312486.
28. Heiderscheit A, Madson A. Use of the iso principle as a central method in mood management: a music psychotherapy case study. Augsburg University Faculty Authored Articles. 2015. https://idun.augsburg.edu/cgi/viewcontent.cgi?article=1046&context=faculty_scholarship
29. Ansdell G. Where performing helps: processes and affordances of performance in community music therapy. In: Stige B, Ansdell G, Elefant C, Pavlicevic M, editors. Where music helps: community music therapy in action and reflection. Aldershot: Ashgate; 2010. p. 161–88.
30. Baker FA. Front and center stage: participants performing songs created during music therapy. Arts Psychother. 2013;40:20–8. https://doi.org/10.1016/j.aip.2012.09.004.
31. National Endowment for the Arts. Creative forces: NEA military healing arts network. Creative forces: clinical therapy. n.d. https://www.arts.gov/initiatives/creative-forces/clinical-therapy

Conclusions and Future Directions

17

Kerry Devlin, Kyurim Kang, and Alexander Pantelyat

Because different people have different music preferences based on their prior experience and cultural background, one size does not fit all when it comes to Music-Based Interventions (MBI): music deemed sublime by one person may be judged as cacophonic by another. There is an inherent tension between the appreciation of music for its aesthetic qualities and the use of music as a treatment and/or invitation to enter a therapeutic process for patients living with neurological diagnoses. Context and intent (e.g., performance for aesthetic appreciation alone vs. music played for therapeutic purposes) are crucial when making the distinction between an artist's and a music therapist's rendering of the same song. However, it is important to acknowledge that a patient may perceive the former as therapeutic and the latter as having aesthetic qualities.

This book describes the diverse applications of music for neurological diagnoses. Achieving a comprehensive understanding of the effectiveness of music therapy necessitates an exploration of the underlying brain mechanisms in conjunction with individuals' behavioral patterns, functional capabilities, and real-life experiences. This holistic approach would not only enhance our understanding of the therapeutic impact of music, but also equip music therapists and other healthcare professionals with the means to deliver more personalized therapy to individuals.

K. Devlin · K. Kang · A. Pantelyat (✉)
Johns Hopkins Center for Music and Medicine, Department of Neurology,
Johns Hopkins University School of Medicine, Baltimore, MD, USA
e-mail: kdevlin5@jh.edu; kkang19@jhmi.edu; apantel1@jhmi.edu

© The Author(s), under exclusive license to Springer Nature Switzerland AG 2023
K. Devlin et al. (eds.), *Music Therapy and Music-Based Interventions in Neurology*, Current Clinical Neurology,
https://doi.org/10.1007/978-3-031-47092-9_17

Considerations Related to Mechanisms for MBI

Music perception and cognition mechanisms have been extensively investigated as mentioned in Chap. 2 (*Mechanisms of Music Therapy and Music-Based Interventions*), yet there remains a notable gap in the understanding of the neural mechanisms underlying the therapeutic effects of music therapy [1]. It is particularly essential to understand the brain mechanisms involved in music-based interventions and music therapy in the context of rehabilitation. In Chaps. 3 (*Music for Stroke Rehabilitation*), 4 (*Music for Traumatic Brain Injury and Impaired Consciousness*), 5 (*Music for Movement Disorders*), 6 (*Music for Speech Disorders*), and 7 (*Music for Memory Disorders*), the authors emphasized the connection between music and neural plasticity. A better understanding of this will facilitate the development of evidence-based interventions that target specific areas of impairment, promote neural plasticity, and enhance functional recovery.

Not only is it crucial to explore the effectiveness of music therapy through pre- and post-session assessments, but it is equally important to comprehend the intricate processes that occur *during* music therapy sessions, including neuro-physiological and behavioral responses between therapists and clients (e.g., measuring brain activity using EEG during the NMT® sessions, as indicated in Chap. 5 (*Music for Movement Disorders*). Although there are limitations to employing brain imaging techniques during these sessions, primarily due to issues of portability and complexity, it is imperative to further investigate the brain mechanisms within ecologically valid environments.

Considerations for Co-constructing MBI in Context

This text illustrates the importance of considering patients living with neurologic diagnoses *in context* – in the context of their illness progression and prognosis, their culture, their identities, their lived experiences, their own reasons for seeking out music as a therapeutic modality, and the dynamics of the systems patients navigate to access care. We call for a reconceptualization of MBI to be a dynamic unfolding process between patient and therapist that values the contributions and expertise *each* has to offer. Consider Chap. 15 (*Music Therapy and Music-Based Approaches with Autistic People: A Neurodiversity Paradigm-Informed Perspective*), for example, which emphasizes the importance of moving away from a deficit-oriented understanding of autism in favor of embracing the collective desires and rights of autistic people to determine what *they* want from their own care – rather than what the therapist thinks they should want.

Given this, we consider MBI selection and implementation as a co-constructed, nuanced process that centers the therapist's knowledge concurrently with the patients' own knowledge and preferences. Some questions to consider when utilizing MBI in collaborative fashion may include (but are not limited to):

1. Is the therapist/music practitioner unintentionally prioritizing one way of working over others, and is this prioritization actually in service of the patient? Where might opportunities exist to shift in and out of different ways of working in response to the patient's needs? *Consider the virtual support group in Chap. 5 (Music for Movement Disorders), which transparently incorporates multiple approaches to music therapy practice in response to the group's interests, needs, and desires, and the merits of exploring "how disparate approaches can co-exist to provide nuanced care" articulated in Chap. 9 (Music Therapy and Music-Based Interventions for Neurologic Palliative Care).*
2. Has the patient indicated a desire for functional improvement or to have an aspect of their symptomology "fixed"? Are they experiencing a health process like active death that simply cannot be changed? Or both? *Chapters 5 (Music for Movement Disorders), 7 (Music for Memory Disorders), 8 (Music for Neuro-Oncological Disorders), 9 (Music Therapy and Music-Based Interventions for Neurologic Palliative Care), 10 (Music for Autoimmune Neurological Disorders), 15 (Music Therapy and Music-Based Approaches with Autistic People: A Neurodiversity Paradigm-Informed Perspective), and 16 (Psychosocial Aspects of Music Therapy), among others, invite us to consider our role as music therapists and practitioners in non-binary fashion, acknowledging that different patients have different preferences – and that disability pride can still co-exist with a desire to experience symptom reduction.*
3. How do the patient's and therapist's own identities and experiences with neurologic illness define the ways in which they think about constructs like health, therapy, and music? *See Chap. 10 (Music for Autoimmune Neurological Disorders) discussion of the importance of intersectional considerations in work with patients living with chronic diagnoses like MS as one such example.*
4. In what ways are the patient's cultural, musical, and personal preferences realized in MBI selection, or are there barriers that prevent individualization? Who determines what music is "good" or "clinically relevant", and why? *Chapter 2 (Mechanisms of Music Therapy and Music-Based Interventions), for example, highlights the fact that the literature tends to prioritize study and integration of Western music in healthcare – excluding cross-cultural perspectives. Consider also the balance between what the literature indicates as "most effective" music in specific contexts, such as the "Mozart effect" discussed in Chap. 11 (Music for Epilepsy), versus the importance of patient-selected music for reducing pre-operative anxiety and postoperative pain, as indicated in Chap. 12 (Music for Surgical/Perioperative Care).*
5. Would the patient benefit from work that takes place within a long-term therapeutic relationship to best support their changing needs and desires over time? Is this clear at the time of initial assessment, or could this be emergent from initial engagement in more outcome-oriented therapeutic processes? *John's Journey in Chap. 16 (Psychosocial Aspects of Music Therapy) illustrates how longitudinal work can shift from "skills-oriented goals" to emphasizing personhood, agency, and grief processing in response to what was most meaningful to John at*

different points in time. Chapter 5 (Music for Movement Disorders) also emphasizes utilizing NMT® *to support emotional and social wellbeing.*

6. In what ways can or should opportunities for practice outside the therapy room be incorporated into a patient's daily routines, and what tools are needed to make this feasible and safe? *Therapeutic technology covered in Chap. 14 like Spiritune®, MedRhythms® and other home-based applications/wearables represent new horizons (and potentially, new safety considerations) for use of personalized MBI access available to patients in their homes. In addition, telehealth sessions (discussed in Chap. 13) continue to develop in ways that demonstrate great potential for further expansion beyond the context of COVID-19 to promote home-based therapy access.*

Parting Thoughts

We believe it is important to highlight that this book does not include patient voices, which are crucial to understanding the impact and experience of MBI in the context of their own lives. It is ultimately the responsibility of clinicians to consider the extent to which patient voices are positioned in the center of MBI selection and delivery. This should be explored in parallel with questions of power, equity, agency, flexibility, and cultural context – all essential considerations raised by many chapter authors in this book. We hope that future versions of this book – and other writings in this area – will embody the spirit of collaboration by positioning patients as co-authors and/or contributors (per their comfort level). This would ensure that patients' lived experiences are amplified in equitable fashion alongside those of music therapists, practitioners, and medical professionals.

By embracing the multifaceted nature of music, we can bridge the gap between science, medicine, and the arts, ultimately paving the way for improved neurological care and enhanced well-being for individuals living with neurological diagnoses.

Reference

1. Chen WG, Iversen JR, Kao MH, Loui P, Patel AD, Zatorre RJ, et al. Music and brain circuitry: strategies for strengthening evidence-based research for music-based interventions. J Neurosci. 2022;42(45):8498–507.

Appendices

Appendix A: Assessment Tool Examples

Neuroimaging Techniques	Description
Magnetic Resonance Imaging (MRI)	Noninvasively evaluates structural (volumetric, diffusion-weighted) and functional (resting state, task-based) brain anatomy and activity
Functional magnetic resonance imaging (BOLD fMRI)	Measures oxygenated blood flow changes to highlight areas of activation at rest and during specified actions
Electroencephalography (EEG)	Measures electrical activity of the brain via electrodes placed on the scalp
Diffusion Weighted Imaging (DWI)	Maps how water travels along white matter tracts by using water molecules as contrast in MR images
Magnetoencephalography (MEG)	Measures neuronal electrical activity to provide timing and spatial information about brain activity
Functional near-infrared spectroscopy (fNIRS)	Measures changes in blood oxygenation and hemodynamics in the brain of one or multiple individuals. Does not require lying still (unlike MRI).
Positron Emission Tomography (PET)	Quantifies physiological processes by measuring the concentration of the radiopharmaceutical at specific locations in the body.

Appendix B: Auditory Entrainment and Oscillatory Coupling as Potential Neural Mechanisms for Rhythm-Based Interventions (Chap. 5)

Entrainment is defined as the alignment of one or more oscillating systems to the periodicity of an external rhythm, whereby the external rhythm unidirectionally influences the oscillating system(s) [1]. External auditory rhythm can entrain brain oscillations whose frequency is the closest match to the temporal structure of the inputs [1, 2] and frequency of external auditory rhythm ranging between 1 Hz and 4 Hz (delta frequency) can strongly induce entrainment in the auditory cortex [3–6]. For example, a pure tone with an interstimulus interval of 767 ms rapidly entrained ongoing EEG delta waves in the primary auditory cortex [7]. Another EEG study further showed that listening to periodic auditory stimuli between 1 and 5 Hz induced delta entrainment (i.e., phase locking measured using stimulus-locked inter-trial coherence) maximizing at 2 Hz that corresponds with the optimal tempo identified during repetitive sensorimotor behavior [8].

In addition, slow (1–5 Hz) frequency oscillations can recruit a larger number of neurons in a broad area of the brain [2]. This may allow for the auditory rhythm with delta frequency to couple or synchronize delta oscillations between auditory and non-auditory areas of the brain [9–12]. Slow frequency oscillations can also couple faster frequency oscillations [3, 12–14]. In the aforementioned EEG study, listening to the repetitive auditory stimuli also induced phase coupling in beta/gamma in which the augmented gamma synchronization appeared to be mediated by increased beta activity [8].

The beta band ranges between 13 Hz and 25–30 Hz and is thought to be predominant and specific rhythm during rest in the motor cortex and associated with action planning and execution [10]. Beta oscillatory power suppression appears preceding movement in self-paced motor tasks [15, 16] and following warning and imperative cues [17, 18]. It also facilitates a new movement [19] and required actions [17, 18, 20]. In addition, beta band oscillation is thought to reflect more internal top-down processes such as temporal attention [21, 22], translation of timing information to auditory-motor coordination [23], and anticipation/prediction of timing/following events [3, 4, 24, 25]. The anticipation of timing is thought to play an important role in the beneficial effects of RAS® [26].

Whether clinical populations with movement disorders also show the auditory entrainment and oscillatory coupling remains to be fully elucidated. In one study during finger tapping with RAS® of interstimulus interval of 1000 ms, people with PD showed intact auditory entrainment in beta and gamma frequency ranges measured by phase-locking [27]. Another study also demonstrated that individuals with PD showed auditory entrainment in alpha and beta frequency bands, which was similar relative to healthy individuals [28]. However, this study also reported that the effects of auditory beta entrainment did not transfer to the motor area in the PD group, showing reduced beta modulation depth compared to the healthy control group. Nonetheless, there were no significant differences in task performance between the two groups. Discrepancies in the findings between the two studies may be due to the differences in the experimental task, outcome measure of entrainment, recording area of the brain, and disease characteristics of PD participants.

References

1. Lakatos P, Gross J, Thut G. A new unifying account of the roles of neuronal entrainment. Current Biology. 2019;29(18):R890–905.
2. Buzsaki G. Rhythms of the Brain. Oxford University Press; 2006.
3. Lakatos P, Shah AS, Knuth KH, Ulbert I, Karmos G, Schroeder CE. An oscillatory hierarchy controlling neuronal excitability and stimulus processing in the auditory cortex. J Neurophysiol. 2005;94(3):1904–11.
4. Nobre AC, Van Ede F. Anticipated moments: temporal structure in attention. Nat Rev Neurosci. 2018;19(1):34–48.

5. Nozaradan S, Peretz I, Missal M, Mouraux A. Tagging the neuronal entrainment to beat and meter. J Neurosci. 2011;31(28):10234–40.
6. Thaut MH, Miller RA, Schauer LM. Multiple synchronization strategies in rhythmic sensorimotor tasks: phase vs period correction. Biol Cybern. 1998;79(3):241–50.
7. Lakatos P, Pincze Z, Fu KMG, Javitt DC, Karmos G, Schroeder CE. Timing of pure tone and noise-evoked responses in macaque auditory cortex. Neuroreport. 2005;16(9):933–7.
8. Will U, Berg E. Brain wave synchronization and entrainment to periodic acoustic stimuli. Neurosci Lett. 2007;424(1):55–60.
9. Canolty RT, Knight RT. The functional role of cross-frequency coupling. Trends Cogn Sci. 2010;14(11):506–15.
10. Morillon B, Arnal LH, Schroeder CE, Keitel A. Prominence of delta oscillatory rhythms in the motor cortex and their relevance for auditory and speech perception. Neurosci Biobehav Rev. 2019;107:136–42.
11. Peelle JE, Davis MH. Neural oscillations carry speech rhythm through to comprehension. Front Psychol. 2012;3:320.
12. Saleh M, Reimer J, Penn R, Ojakangas CL, Hatsopoulos NG. Fast and slow oscillations in human primary motor cortex predict oncoming behaviorally relevant cues. Neuron. 2010;65(4):461–71.
13. Escoffier N, Herrmann CS, Schirmer A. Auditory rhythms entrain visual processes in the human brain: evidence from evoked oscillations and event-related potentials. NeuroImage. 2015;111:267–76.
14. Fujioka T, Trainor LJ, Large EW, Ross B. Internalized timing of isochronous sounds is represented in neuromagnetic beta oscillations. J Neurosci. 2012;32(5):1791–802.
15. Hari R, Salmelin R. Human cortical oscillations: a neuromagnetic view through the skull. Trends Neurosci. 1997;20(1):44–9.
16. Pfurtscheller G, Da Silva FL. Event-related EEG/MEG synchronization and desynchronization: basic principles. Clin Neurophysiol. 1999;110(11):1842–57.
17. Doyle LM, Yarrow K, Brown P. Lateralization of event-related beta desynchronization in the EEG during pre-cued reaction time tasks. Clin Neurophysiol. 2005;116(8):1879–88.
18. Williams D, KuÈhn A, Kupsch A, Tijssen M, van Bruggen G, Speelman H, et al. Behavioural cues are associated with modulations of synchronous oscillations in the human subthalamic nucleus. Brain. 2003;126(9):1975–85.
19. Gilbertson T, Lalo E, Doyle L, Di Lazzaro V, Cioni B, Brown P. Existing motor state is favored at the expense of new movement during 13–35 Hz oscillatory synchrony in the human corticospinal system. J Neurosci. 2005;25(34):7771–9.
20. Kühn AA, Williams D, Kupsch A, Limousin P, Hariz M, Schneider GH, et al. Event-related beta desynchronization in human subthalamic nucleus correlates with motor performance. Brain. 2004;127(4):735–46.
21. Alanazi FI, Al-Ozzi TM, Kalia SK, Hodaie M, Lozano AM, Cohn M, et al. Neurophysiological responses of globus pallidus internus during the auditory oddball task in Parkinson's disease. Neurobiol Dis. 2021;159:105490.

22. Stefanics G, Hangya B, Hernádi I, Winkler I, Lakatos P, Ulbert I. Phase entrainment of human delta oscillations can mediate the effects of expectation on reaction speed. J Neurosc. 2010;30(41):13578–85.
23. Fujioka T, Ross B, Trainor LJ. Beta-band oscillations represent auditory beat and its metrical hierarchy in perception and imagery. Journal of Neuroscience. 2015;35(45):15187–98.
24. Crasta JE, Thaut MH, Anderson CW, Davies PL, Gavin WJ. Auditory priming improves neural synchronization in auditory-motor entrainment. Neuropsychologia. 2018;117:102–12.
25. Hickok G, Farahbod H, Saberi K. The rhythm of perception: entrainment to acoustic rhythms induces subsequent perceptual oscillation. Psychol Sci. 2015;26(7):1006–13.
26. Thaut MH. The discovery of human auditory–motor entrainment and its role in the development of neurologic music therapy. Prog Brain Res. 2015;217:253–66.
27. Buard I, Dewispelaere WB, Teale P, Rojas DC, Kronberg E, Thaut MH, et al. Auditory entrainment of motor responses in older adults with and without Parkinson's disease: an MEG study. Neurosci Lett. 2019;708:134331.
28. Te Woerd ES, Oostenveld R, de Lange FP, Praamstra P. Impaired auditory-to-motor entrainment in Parkinson's disease. J Neurophysiol. 2017;117(5):1853–64.

Appendix C: Rhythmic Auditory Cueing Studies in Parkinson Disease with Assessments of Brain Function (Chap. 5)

Appendices

Study design/techniques	Participants/groups/characteristics[a]	Task or intervention	Primary outcome measures	Primary results (with rhythmic auditory cueing)
Experimental				
LFPs [1]	16 PD on med with DBS electrode bilaterally implanted in STN (age 61 ± 4 years; DD, 12 ± 4 years, 1 left-handed, 1 woman)	Experimental tasks: Synchronize steps, watching each heel strike displayed on the video (42 s containing 21 left and 21 right heel strikes) with and without auditory cueing (metronome sound at interval of 1's) Active control task: Watch the video without moving and to think of anything unrelated to walking such as holidays or upcoming plans	Step-to-cue difference, step interval durations, step timing variability, and step interval variability Beta power modulation	Significant reduction in step timing variability, higher beta modulation, and higher beta rebound and faster beta decrease associated with longer interval of steps
LFPs [2]	7 PD on med with DBS electrode bilaterally implanted in STN (mean age 63 ± 6.5 years; DD range, 6–18 years; 3 women)	Move one hand or foot between two dots separated by 30 cm, cued by metronome of 1.6, 3.2, and 4.8 Hz during normal DBS, 80% DBS amplitude, or no DBS	Action tremor	With auditory cueing and DBS combined, significant reduction in the number of extremities showing action tremor

Study design/ techniques	Participants/groups/ characteristics[a]	Task or intervention	Primary outcome measures	Primary results (with rhythmic auditory cueing)
MEG [3]	21 PD on med (age 66.3 ± 7.4 years, 12 women) 23 HCs (age 65.8 ± 7.7 years, 12 women)	Right index finger tapping with rhythmic (every 1 s) and non-rhythmic (every 1 s ± 200 ms) acoustic burst stimuli (a total of 6 sequences of 30 s separated by a 5-s rest period)	Mean distance to cue (average time difference between onset of cue and response, weighted by the relative number of pre- and post-cue responses) Oscillatory neuronal activity within 15–80 Hz	No significant group difference in mean distance to cue or the number of pre- and post-cue responses During the rhythmic condition, greater activation in bilateral frontal, supplementary and primary motor areas, and occipital regions in HC compared to PD and greater activity in bilateral angular and supramarginal gyri, and left rolandic operculum, in PD compared to HC During rhythmic condition compared to non-rhythmic condition, greater activity in temporal, parietal, and motor areas in HC, and greater activity in left angular and right middle frontal gyri in PD
fMRI [4]	27 PD on med who could walk independently and showed improvement of gait with RAS® (age, 74.6 ± 6.8 years; 16 women; DD, 6.3 ± 4.5 years; H&Y II:III (n), 11:16; MMSE-Japanese, 27.6 ± 2.3) 25 HC (age 71.8 ± 5.5 years, 12 women; MMSE-J 27.9 ± 2.2)	Experimental tasks: Gait imagery with white noise and with rhythmic stimuli (electrically synthesized beep sound presented at 100 bpm) Control tasks: rest with white noise and with rhythmic stimuli	Brain activation	Rhythmic stimuli vs white noise: reduced activity in the left parietal operculum in PD and no difference between the two conditions in HC Gait imaginary vs rest: no difference or interaction in PD and increased activity in the left superior temporal gyrus with rhythmic stimuli compared with white noise in HC

fMRI [5]	14 PD on med (age range, 51–81; 5 women) 13 HC (age range, 51–78; 8 women)	1 Hz finger tapping with/without rhythmic auditory cues 4 Hz finger tapping with/without rhythmic auditory cues	Signed error (difference in timing between tap and onset of the stimulus tone) Behavioral freezing (presence of large pauses during tapping) Intra- and inter-network connectivity	Significant reduction in variability of finger tapping across the group with rhythmic auditory cues With 4 Hz rhythmic auditory cues, significant reduction in freezing in PD No group difference in signed error During 1 Hz tapping, greater intra-network connectivity in auditory, salience, motor/IFG, basal ganglia/thalamus, and motor/cerebellar networks in HC compared to PD Greater inter-network connectivity in three pairs of networks: between auditory and executive control networks; between executive control and motor/cerebellar network; and between auditory and visual networks in PD compared to HC with and without RAS®
[11C]-DTBZ PET [6]	28 PD (age, 67.3 ± 7.5 years; 5 women; DD, 5.2 ± 3.2 years; H&Y I–III; UPDRS motor: 21.6 ± 7.2 (off), 19.7 ± 7.2 (on); MoCA: 25.9 ± 3.1 (off), 26.3 ± 2.7 (on); MMSE: 28.9 ± 1.1 (off), 28.8 ± 1.2 (on)	Synchronized right or left index finger tapping to tone sequences of either 500 ms, 1000 ms or 1500 ms time intervals while ON levodopa placebo pill	Accuracy and variability of synchronization Striatal dopaminergic denervation	Less accurate synchronizing to 500 ms time interval, compared to the 1000 ms and 1500 ms time intervals No effects of medication state or hand used on synchronization accuracy or variability No association between dopaminergic denervation and synchronization accuracy or variability 3 subgroups (based on degree of denervation) had partially paralleled behavioral differences in synchronization accuracy

Study design/techniques	Participants/groups/characteristics[a]	Task or intervention	Primary outcome measures	Primary results (with rhythmic auditory cueing)
Intervention				
RCT with EEG [7]	50 PD on med: RAS® (n = 25; age, 70 ± 8 years; 9 women; DD, 10 ± 3 years; H&Y, 3 ± 1; MMSE, 26 ± 3) No RAS® (n = 25; age, 73 ± 8 years; 9 women; DD, 9.3 ± 3 years; H&Y, 3 ± 1; MMSE, 25 ± 3)	Individual treadmill gait training with or without RAS® supervised by physiotherapists (30 minutes, once/day, 5 days/week for 8 weeks) Walk along with beat (superimposed salient high-pitch bell sound) of music "angel elsewhere" presented with the lyrics, which reaches a target music tempo of ~120 bpm	FGA; UPDRS; BBS; FES; 10MWT; TUG; GQI ERD/ERS magnitude and functional connectivity in alpha and beta bands	Significant improvement of FES, FGA, and UPDRS and increase in GQI in RAS® group Significant improvement of BBS and TUG in both groups Greater EEG power increase in ERD and ERS in alpha and beta bands within the frontal and centro-parietal areas in RAS® group Greater fronto-centroparietal/temporal connectivity in beta band in RAS® group Greater increase in the fronto-centroparietal and fronto-temporal beta connectivity, correlated with greater FGA improvement in both groups combined
RCT with EEG [8][b]	50 PD on med RAS® (n = 25; age, 70 ± 8 years; 9 women; DD, 10 ± 3 years; H&Y, 3 ± 1; MMSE, 26 ± 3) No RAS® (n = 25; age 73 ± 8 years; 9 women; DD, 9.3 ± 3 years; H&Y, 3 ± 1; MMSE, 25 ± 3)	Individual treadmill gait training with or without RAS® supervised by physiotherapists (30 min, once/day, 5 days/week for 8 weeks) Walk along with beat (superimposed salient high-pitch bell sound) of music "Animals Everywhere" presented with the lyrics, which reaches a target music tempo of ~120 bpm	EEG amplitude/activation	Greater reduction in frontal area activation and more potentiation in activation of the centroparietal areas in RAS® compared to No RAS® Gait-cycle specificity and reduction of cerebellar activity observed only in RAS Improvement of gait (combined FGA and GQI) associated with decreased activity within frontal and cerebellar regions, and with increased activity within central and parietal regions in both groups combined

Case-control study with EEG [9]	20 PD on med DBS (n = 10; age, 62 ± 5 years; 4 women; DD, 15 ± 2 years; H&Y, 3 (2.5–3); MMSE, 26 (25–27); UPDRS-III ON, 17 (14–21); UPDRS III OFF, 32 (27–44) No DBS (n = 10; age, 62 ± 4 years; 5 women; DD, 14 ± 2 years; H&Y, 2.5 (2.5–2.9); MMSE, 27 (27); UPDRS-III ON, 28 (27,28); UPDRS III OFF, 37 (34–45)	Treadmill gait training with or without RAS® on med (30 min, once/day, 4 days/week for a month). Simple two-accent metronome sounds of 120 bpm or the maximum tolerable bpm	UPDRS-III ON and OFF; BBS; FES; 10MWT; TUG; ACE-R ERSP in relation to RAS® provision in the alpha and beta frequency ranges	Significant improvement in BBS, FES, and ACE-R in both groups Greater improvement in UPDRS OFF, TUG, and 10 MWT in DBS group Greater remodulation of sensorimotor alpha and beta oscillations associated with the gait cycle in both groups Significant association between beta power percent change (decrease) within motor programing ROI and 10 MWT percent change (increase) in both groups combined
Case study with MEG [10]	3 PD on med: Participant 1: 62 yo, woman; H&Y, 2.5; UPDRS III, 23 Participant 2: 70 yo, man, H&Y, 3; UPDRS III, 52 Participant 3: 72 yo, H&Y, 2; UPDRS III, 24	15 sessions of somatosensory-related NMT® techniques (3 times/week for 5 consecutive weeks) provided by an NMT-certified music therapist Each session consisting of bimanual exercises using a keyboard, castanets and miscellaneous objects to strengthen fine motor muscles Each finger movement cued by either a metronome or beats produced by the therapist playing musical instrument	UPDRS-III; GPT; Finger-Thumb Opposition Task Evoked power and functional connectivity of frequency ranging 15–80 Hz in left auditory and primary motor cortices during cued finger tapping	Improvement in one or more areas of fine motor functions for either dominant hand or both hands in all participants Possibly coinciding increase in beta-range evoked power Increased functional connectivity between primary auditory and motor cortices

Study design/techniques	Participants/groups/characteristics[a]	Task or intervention	Primary outcome measures	Primary results (with rhythmic auditory cueing)
CCT with [18F]-FDG PET [11]	9 PD on med (age, 61 ± 5; 4 women; DD range, 3–8 years; H&Y range, 1–2.5; UPDRS range, 12–45) 5 HCs (age, 63 ± 4; 3 women)	Gait and repetitive arm training consisting of 20 1-h sessions over 4 weeks Gait and/or upper limb movements cued by a metronome at rates of 30–150 bpm Upper limb movements cued by a metronome at rates of 0.5 and 4 Hz	Gait parameters (velocity; step length; cadence; variability) Finger tapping (frequency; variability) Glucose metabolism	Higher variability in gait and finger tapping and less velocity in PD compared to HC before training In PD, significantly decrease in variability of gait and finger tapping to the level of HC after training Hypometabolism in right parietal and temporal lobes, left temporal lobe, left frontal lobe, and left cerebellum (culmen of anterior lobe) in PD compared to HC before training. Increase in glucose metabolism in right cerebellum including anterior lobe and dentate nucleus, as well as right parietal (BA39) and temporal lobes after training compared to before training in PD

ACE–R Addenbrooke's cognitive examination–revised, BA Brodmann area, BBS Berg balance scale, bpm beat per minute, CCT controlled clinical trial, DBS deep brain stimulation, DD duration of disease, EEG electroencephalogram, ERD event-related desynchronization, ERS event-related synchronization, ERSP event-related spectral perturbation, FES falls efficacy scale, FGA functional gait assessment, fMRI functional magnetic resonance imaging, GPT grooved pegboard test, GQI gait quality index, HC healthy control, H&Y Hoehn and Yahr, LFP local field potential, MEG magnetoencephalography, MMSE mini mental state examination, MoCA Montreal cognitive assessment, NMT neurologic music therapy, PET positron emission tomography, PD Parkinson's disease, RAS® rhythmic auditory stimulation, RCT randomized control trial, ROI region of interest, STN subthalamic nucleus, TUG timed up-and-go test, yo years old, 10MWT 10-m walking test, UPDRS unified Parkinson's disease rating scale

[a]Numbers presented as mean ± SD or median (interquartile range) unless stated otherwise
[b]Secondary EEG analysis of the study by Naro et al. [9]

References

1. Fischer P, Chen CC, Chang YJ, et al. Alternating modulation of subthalamic nucleus beta oscillations during stepping. J Neurosci. 2018;38(22):5111–21.
2. Heida T, Wentink EC, Zhao Y, Marani E. Effects of STN DBS and auditory cueing on the performance of sequential movements and the occurrence of action tremor in Parkinson's disease. J Neuroeng Rehabil. 2014;11:135.
3. Buard I, Dewispelaere WB, Teale P, et al. Auditory entrainment of motor responses in older adults with and without Parkinson's disease: an MEG study. Neurosci Lett. 2019;708:134331.
4. Nishida D, Mizuno K, Yamada E, Hanakawa T, Liu M, Tsuji T. The neural correlates of gait improvement by rhythmic sound stimulation in adults with Parkinson's disease – a functional magnetic resonance imaging study. Parkinsonism Relat Disord. 2021;84:91–7.
5. Braunlich K, Seger CA, Jentink KG, Buard I, Kluger BM, Thaut MH. Rhythmic auditory cues shape neural network recruitment in Parkinson's disease during repetitive motor behavior. Eur J Neurosci. 2019;49(6):849–58.
6. Miller NS, Kwak Y, Bohnen NI, Muller ML, Dayalu P, Seidler RD. The pattern of striatal dopaminergic denervation explains sensorimotor synchronization accuracy in Parkinson's disease. Behav Brain Res. 2013;257:100–10.
7. Calabro RS, Naro A, Filoni S, et al. Walking to your right music: a randomized controlled trial on the novel use of treadmill plus music in Parkinson's disease. J Neuroeng Rehabil. 2019;16(1):68.
8. Naro A, Pignolo L, Bruschetta D, Calabro RS. What about the role of the cerebellum in music-associated functional recovery? A secondary EEG analysis of a randomized clinical trial in patients with Parkinson disease. Parkinsonism Relat Disord. 2022;96:57–64.
9. Naro A, Pignolo L, Sorbera C, Latella D, Billeri L, Manuli A, Portaro S, Bruschetta D, Calabrò RS. A case-controlled pilot study on rhythmic auditory stimulation-assisted gait training and conventional physiotherapy in patients with Parkinson's disease submitted to deep brain stimulation. Front Neurol. 2020;11:794.
10. Buard I, Dewispelaere WB, Thaut M, Kluger BM. Preliminary neurophysiological evidence of altered cortical activity and connectivity with neurologic music therapy in Parkinson's disease. Front Neurosci. 2019;13:105.
11. del Olmo MF, Cudeiro J. Temporal variability of gait in Parkinson disease: effects of a rehabilitation programme based on rhythmic sound cues. Parkinsonism Relat Disord. 2005;11(1):25–33.

Index

A
Action observation/execution matching system, 140
Acute pain, music for, 14
Affect regulation (AR), music for, 15, 16
Alzheimer disease dementia, 5
American Psychological Association, 201
Anti-oppressive music therapy, 187, 192
Antiseizure medication (ASM) therapy, 138
Applied behavioral analysis (ABA), 190
Arousal, 39, 41, 46
Artism Ensemble, 193, 194
Assessment of quality of life (AQoL), 126
Associative memories, 88
Atypical parkinsonian (APD) disorders, 59
Auditory cues, 88
Auditory-motor coupling, 12
Auditory processing, 72
Autism
  behavioral music therapy, 190
  community music therapy, 191
  ethnographic and related approaches, 193
  ethnomusicology, 194–195
  music therapy, 191–194
  neurodiversity, 191–193
Autism Diagnostic Observation Schedule assessment tool (ADOS), 190
Autism symptom severity, 192
Autoimmune disorders, 5
Autoimmune encephalitis (AE), 124, 126, 131
Autoimmune neurological disorder
  assessments, 126
  clinical cases, 130–132
  definition/background, 124, 125
  music therapy, 127
    biopsychosocial approaches, 128
    cognitive dimension, 129
    family/support system dimension, 130
    interventions, 127, 128
    physical dimension, 129
    psychoeducational approach, 128
    psychoemotional dimension, 129
    psychosocial dimension, 129
    spiritual dimension, 130
  patient's intersectionalities, 126, 127
  symptoms, 125
Awake craniotomy, 153

B
Barcelona Music Reward Questionnaire, 27
Behavioral music therapy approaches, 6, 190
Behavioral and psychological symptoms of dementia (BPSD), 164
Body-mind-medicine, 128
Brain cancer, 98
  music therapy in, 99
  radiation therapy for, 101, 102

C
Calm®, 176, 177
Cancer
  chemotherapy and other systemic therapy, 102
  MBI for, 98, 99
    in palliative care of patients, 102, 103
    perspectives on clinical practice and research opportunities, 103, 104
  after surgery, music therapy, 100
Caregivers (CGs), 164
Central pattern generator (CPG), 11
Cerebellar abnormality, 51
Certification Board of Music Therapists (MT-BC), 123, 124
Chronic pain, music for, 15

Cognitive aging, 24
Cognitive dimension, 129
Community choirs, 92
Community music therapy, 191
Comorbid mood disorders, 203
Consciousness, definition of, 39
Continuing telehealth services, 163
"Continuous Ambient Relaxation Environment" (C.A.R.E.), 155
Cortical dynamics theory, 141
Corticobasal syndrome (CBS), 63
COVID-19, 6, 161, 163, 165, 167, 176, 216
Creative Forces®, 200
Culture, environment, and social influences, 202

D
Dementia, 86
Demyelinating disease, 130
Descending/"top-down" pain modulatory system (DPMS), 15
Developmental language disorders (DLD), 74, 78, 79
Digital technology, 174
Disorders of consciousness (DoC), 37, 39
    guidelines for delivery of music interventions in, 45
    music therapy for, 40–42
Dopamine, 13, 51
DSM 5 criteria, 189
DSM-V cultural formulation interview, 126

E
Empathy, 13
Epilepsy
    action observation/execution matching system/human mirror neuron system, 140
    complex and highly organized architecture, 141
    cortical dynamics theory, 141
    definitions and background, 138–140
    1/f theory, 141
    interictal epileptiform discharges, 139–140, 142
    Jobst group perspectives, 138
    Mozart effect, 140, 141
    music
        and epilepsy comorbidities, 145
        and interictal epileptiform discharges, 141–143
        and seizures, 143–145
    Rektor group perspectives, 138
    Trion model, 141
Epileptiform activity, 139
Episodic memories, 88
Ethnomusicology, 194–195
Event-related deysnchronization (ERD), 51
Event-related synchronization (ERS), 51
Exploratory world music playground (E-WoMP), 194

F
Family/support system dimension, 130
Freezing of gait (FoG), 58
1/f resonance mechanism, 141
1/f theory, 141

G
Gait and balance, 11
Glioblastoma, 104
Glioma, 98
Group singing, 28

H
Headspace®, 176, 177
Health Insurance Portability and Accountability Act, 207
Hemiparesis, 11
Home-based music therapy, 31
Hospice care, 110
Human mirror neuron system, 140
Huntington's disease, 131
Hypnotic music, 150
Hypothalamic-pituitary-adrenal axis, 4

I
Ictal activity (seizure activity), 139
Imaging techniques, 50
Impact areas questionnaire, 164
Infant directed speech (IDS), 73
Integrative music therapy approaches, 60–62
Interictal epileptiform discharges (IEDs), 139–143
International Federation Reference Centre for Psychosocial Support, 201
Interpersonal synchrony, music for, 12, 13
Intracranial electroencephalography (iEEG), 138, 142
Intraoperative nursing, 150
Involuntary memories, 87
Iso-principle process, 204

K
Kidscope questionnaire, 101

L
Lab-based music interventions, 175
Language
 definition of, 72
 development, 73
 music and, 74, 75
Large-scale TIME-A trial of music therapy, 190
Live music therapy, 99, 152
Lucid's music curation system, 177

M
Mechanisms of music therapy
 for affect regulation, 15, 16
 for interpersonal synchrony, 12, 13
 for pain management, 14, 15
 in motor rehabilitation, 10–12
Medical model approaches, 189
MedRhythms®, 174, 179, 181, 182
MedRhythms®products, 182
Melodic intonation therapy (MIT), 5, 12, 28, 75, 76, 117
Memory
 definition of, 86
 music as trigger, 89
 musical, 88, 89
Memory disorders
 MBI for, 90
 music therapy for, 91–93
Military Healing Arts Network, 200
Minimal music, 150
Minimalism, *see* Minimal music
Minimally conscious state (MCS), 39
Motor rehabilitation, music in, 10–12
Movement disorders
 definition of, 50
 music-based interventions, 62, 63
 music therapy for
  integrative music therapy approaches, 60–62
  NMT® approaches, 58–60
 rhythm for
  neural mechanisms of auditory rhythm and music, 51
  RAS ®, 51
Mozart effect, 137, 140, 141
Multidisciplinary team, 46
Multiple sclerosis, 123, 124, 127, 130

Music in psychosocial training and counseling (MPC), 165, 202
Music listen, 6, 150–153, 156
Music medicine, 3, 62, 63, 66, 111, 165
Music perception and cognition mechanisms, 214
Music speech stimulation (MUSTIM)®, 117
Music supported therapy (MST), 26, 27
Music therapist vs. ethnomusicologist, 188
Music therapy, 1, 3, 9, 17, 40
 for affect regulation, 15
  for neurological/neurodegenerative conditions, 16
  limbic system, ANS, and HPA axis, 16
 approaches, 151
 clinical pathway, 207
 for interpersonal synchrony, 12, 13
 in motor rehabilitation, 10–12
 for pain management, 14
  for acute pain, 14
  for chronic pain, 15
Music upper limb therapy-integrated (MULT-I), 24, 26, 27, 30, 31
Musical attention control training, 165
Musical Neglect Training®, 26
Musical sonification, 4
Musical speech stimulation, 76
Music-assisted relaxation, 204
Music-based activities, 164
Music-based interventions (MBI), 1, 3, 5–7, 9, 49, 66, 79, 151–153, 156, 157, 213–216
 in cancer care, 98, 99
  in palliative care of patients, 102, 103
  perspectives on clinical practice and research opportunities, 103, 104
 for hospice and palliative care, 113, 114
 for movement disorders, 62, 63
MusicGlove, 4, 28
Music-Play Project (MPP), 194, 195

N
National Association of Social Work, 201
National Endowment for the Arts (NEA), 200
Neurodivergent, 187, 189, 191, 192
Neurodiversity
 autism, 191–193
 definition, 189
 and music therapy, 191, 192
 paradigm, 188, 189, 191, 192, 195
 paradigm-informed approach, 187, 188, 195
 paradigm-informed music therapy, 191, 192

Neurodiversity-affirmative music therapy, 192
Neurodiversity-informed practice, 192
Neurologic music therapy (NMT®) technique, 26, 51, 65, 110, 112, 117, 163, 202
Neurologic music therapy (NMT®) approaches, 58–60, 110, 123, 128, 165
Neuromatrix, 14
Neuro-oncology, 98, 99, 102, 105
Neuropsychiatric inventory assessment, 164
"Neuroqueering" music therapy, 192
Neurorehabilitation, 24, 25, 40
Neurosurgery, 100
Neurotypical, 189, 190, 192
Non-pharmacological interventions, 149
Nordoff-Robbins approach, 27
Nostalgic memories, 87
Nurture authentic communication strategies, 191

O

Oncology, 98
Online music therapy, 163
OPERA hypothesis, 74
Oxytocin, 13

P

Pain management, music for, 14, 15
Palliative care, 109
    intervention and treatment outcome modifications in, 114, 115
    MBI for progressive neurologic disorders in, 113, 114
    music therapy and progressive neurologic disorders in, 111–113
Paradigm shift, 189
Parasympathetic nervous system (PSNS) activity, 152, 156
Parkinson's disease (PD), 49, 50, 52–57, 59, 62, 77
Pathology paradigm, 189, 190, 192
Patient health questionnaire-9 (PHQ-9), 31
Patient's intersectionalities, 126, 127
Patient-selected music, Spotify, 154
Patterned sensory enhancement (PSE), 58, 63, 165
Perioperative music-based interventions, 151–152
Perioperative music therapy approaches, 152–153
Perioperative nursing, 150

Physical dimension, 129
Playing mode vs. stimming mode, 195
Postanesthesia care unit (PACU), 150–156
Post-traumatic amnesia (PTA), 41
    definition of, 39
    music therapy to reduce agitation and increase orientation, 43, 44
Preferred/selected music, 151
Pre-recorded minimalist music, 154
Psychoemotional dimension, 129
Psychosocial aspects
    additional supports, 210
    bio, psycho, and social domains, 201
    check-in, 208
    check-out/discussion and planning, 209–210
    clinical cases, 205–210
    creative engagement time/independent skills practice, 209
    definition, 201
    group description, 208
    music and psychosocial considerations, 202
    music skills review and resourcing, 208
    music therapy
        active music listening, 204
        individuals with neurological disorders, 203
        iso-principle process, 204
        music and imagery, 204
        music-assisted relaxation, 204
        rhythmic entrainment, 204
    music therapy clinical pathway, 207
    performance-based model, 204
    person-centered and strengths-based practices, 201
    psychosocial benefits, 210
    warm-up exercise, 208–209
Psychosocial benefits, 210
Psychosocial dimension, 129

Q

Qualitative analysis, 99
Quality of life (QoL), 98, 102, 105, 109, 111, 118
Quantitative analysis, 99

R

Radiation therapy (RT), 101, 102
Randomized clinical trial (RCT), 41, 99
Receptive music therapy, 41

Index

Reminiscence bump, 87
Resource-oriented approach, 115, 116
Resource-oriented practices, 110
Rhythm processing, 73
Rhythmic auditory stimulation (RAS®), 4, 11, 23, 26, 29, 30, 50, 51, 58, 59, 65
Rhythmic cueing, 52–58
Rhythmic entrainment, 12, 204
Rhythmic sensory stimulation (RSS) treatments, 165
Richards-Campbell Sleep Questionnaire, 100

S

Seizure, 137–140, 143–145
"Sensory Friendly Concerts (SFC)", 191
SingFit PRIME, 180
SingFit STUDIO, 180
SingFit's STUDIO Pro, 180
SingFit-R application, 180
Singing, 28, 75
Social/cultural model approach, 192
Songwriting, 112
Sonification of arm movements, 29
Spatial neglect, music playing to reduce, 27
Speech
 definition of, 72
 development, 73
Speech disorders, 71
 melodic intonation therapy, 75, 76
 music and, 74, 75, 78
Spiritual dimension, 130
Spiritune®, 177, 179, 180, 182
Spontaneous memory, 87
Spotify, 151, 154
State-trait anxiety inventory (STAI) questionnaire, 153–155
Stroke, 23
 definition of, 24
 moving to musical cues, 26
 music listening, 25
 playing musical instruments, 26
  group singing, 28
  MULT-I, 27, 30, 31
  music supported therapy, 27
  singing, 28
  spatial neglect, music playing to reduce, 27
  technology-assisted interventions, 28, 29
  TIMP®, 26

RAS® gait training, 29, 30
 and brain injury, 166–167
Sympathetic nervous system (SNS) activity, 152, 156

T

Technology-assisted interventions
 MusicGlove, 28
 sonification of arm movements, 29
Telehealth
 advantages, 162
 challenges, 163
 definition, 162
 dementia, 164
 emergence of, 163–164
 ethics, 167, 168
 family caregivers, 164
Telehealth music therapy (TMT)
 music medicine approaches, 165
 neurologic music therapy approaches, 165
 stroke and brain injury, 166–167
 supporting persons with PD, 166
 virtual community music experiences, 167
Temozolomide, 103
Therapeutic instrumental music performance (TIMP®), 26, 63, 113
Therapeutic movement experiences (TMEs), 165
Therapeutic singing (TS), 76, 165, 166
Therapeutic technology
 background, 176
 Calm®, 177
 companies and academic labs, 178–179, 182
 Headspace®, 176, 177
 interventions, 175
 lab-based music interventions, 175
 MBI, 174, 176
 MedRhythms, 181
 prescription digital therapeutics, 176
 SingFit, 180
 Spiritune®, 177
 VeraPro®, 180
 Vera™, 180
Therapy for change, 191
TIMP® combined with metronome-cued motor imagery (TIMP®-cMI), 26
TIMP® combined with non-cued motor imagery (TIMP®-MI), 26
Traditional medical model/consensus model approach, 187

Traumatic brain injury (TBI), 4, 37
    definition of, 38
    guidelines for delivery of music interventions in, 45
    music therapy for, 42
    music therapy to stimulate responses in early consciousness rehabilitation, 42, 43
    sequelae and rehabilitation approaches, 39, 40
    severity of, 38
Trion model, 141

U
Unresponsive wakefulness syndrome (UWS), 39

V
Valuing neurodiversity, 191
VeraPro®, 179, 180
Vera™, 179, 180
Vibroacoustic therapy (VAT), 165, 166
Virtual community music experiences, 167
Virtual support group, 64, 65
Visual analogue scales (VAS), 126
Vocal intonation therapy (VIT®), 76, 165

W
Walking-based products, 181
Westmead PTA Scale, 39